ATMOSPHERE, CLOUDS, AND CLIMATE

ATMOSPHERE, CLOUDS, AND CLIMATE

David Randall

PRINCETON UNIVERSITY PRESS *Princeton & Oxford*

Copyright © 2012 by Princeton University Press
Published by Princeton University Press,
41 William Street, Princeton, New Jersey 08540
In the United Kingdom: Princeton University Press,
6 Oxford Street, Woodstock, Oxfordshire OX20 1TW

press.princeton.edu

Library of Congress Cataloging-in-Publication Data

Randall, David A. (David Allan), 1948–
Atmosphere, clouds, and climate / David A. Randall.
p. cm. — (Princeton primers in climate)
Includes bibliographical references and index.
ISBN 978-0-691-14374-3 (hardcover) — ISBN 978-0-691-14375-0 (pbk.)
1. Climatology. 2. Clouds. 3. Atmospheric circulation. I. Title.
QC981.R36 2012
551.6—dc23
2011047912

British Library Cataloging-in-Publication Data is available

This book has been composed in Minion Pro and Avenir

Printed on acid-free paper. ∞

Printed in the United States of America

10 9 8 7 6 5 4 3 2 1

Contents

Preface

....

THIS IS A PRIMER. IT IS AN ATTEMPT TO SKETCH, WITH A few strokes, the role of atmospheric processes in climate—a massive, beautiful, and rapidly advancing subject, full of elegant ideas and amazing facts. My goals are to teach you something about the role of atmospheric processes in climate and to entice you to want to know more.

The book is aimed at college undergraduates who have an interest in climate and some familiarity with basic physics. No background in atmospheric science is assumed. The physical processes that arise in the discussion include radiative transfer, fluid dynamics, and thermodynamics. They are key to a wide range of climate-related topics, including monsoons, winter storms, the water cycle, and anthropogenic climate change.

Atmospheric physics is a highly quantitative subject, so this book contains lots of numbers (mostly in the form of plots), and there are a few equations in almost every chapter. Familiarity with basic calculus is assumed, but there are no complicated derivations. The penalty paid for this simplicity is that the explanations given are much less complete and rigorous than they could be in a more technical book. Chapters 3 and 4 are supplemented with appendixes, to provide a bit more depth for those who are both prepared and interested.

....

The suggestions for further reading have been selected for their accessibility. Additional, more technical references are listed at the back of the book.

Drafts of the manuscript were insightfully critiqued by Mark Branson, Brian Jones, Barbara Whitten, Wayne Schubert, and Richard Somerville. Mark also created the plots used in many of the figures and helped compile the list of terms for the glossary. Louis Uccellini and Richard Grumm, of the National Centers for Environmental Prediction, generously provided help in accessing data needed for one of the figures. Chris Kummerow pointed out what would have been an embarrassing error. Scott Denning and Marcia Donnelson provided encouraging words as the writing neared an end. Last but not least, my wife, Mary Kay, coaxed me on with a promise of bear claws when the book was done. And they were really good.

Princeton University Press conceived and organized the creation of the *Princeton Primers in Climate*—a great idea. It has been a real pleasure to work with Ingrid Gnerlich, Alison Kalett, Kelly Malloy, and Christopher Chung. Ingrid pushed just hard enough, and I thank her for that.

ATMOSPHERE, CLOUDS, AND CLIMATE

1 BASICS

WHAT IS CLIMATE?

Robert Heinlein wrote that "[c]limate is what we expect, weather is what we get."[1] What we expect is *typical* weather, but the weather at a given place, on a given day, can be very atypical.

A dictionary definition of climate is "the average course or condition of the weather at a place usually over a period of years as exhibited by temperature, wind velocity, and precipitation" (Merriam-Webster online, http://www.merriam-webster.com/dictionary/climate). Here "average" refers to a time average. In many reference works, climatological averages are defined to be taken over 30 years, a definition that probably has to do more with the human life span than with any physical time scale.

Simple 30-year averages are not enough because they hide important variability, such as differences between summer and winter. Because of the great importance of the seasonal cycle, climatological averages are often specified for particular months of the year; we might discuss the climatological average precipitation rate for New York City, averaged over 30 Julys, or 30 Januarys.

Climate varies geographically, most obviously between the tropics and the poles. There are also important climate variations with longitude, at a given latitude.[2] For example, the Sahara Desert and the jungles of southeast Asia have very different climates, even though they are at the same latitude. Climate also varies strongly with surface elevation; near the surface, the temperature and water vapor concentration of the air generally decrease upward, and the wind speed often increases upward.

The climate of the Earth as a whole changes with time, for example when ice ages come and go. In fact, the climate of the Earth as a whole is changing right now due to rapid, anthropogenically produced changes in the composition of the atmosphere.

The climate parameters of greatest importance to people are precipitation and temperature. The next few items on the list would include wind speed and direction, and cloudiness. Scientists are interested in a much longer list of parameters, of course.

The "average course or condition of the weather," mentioned in the dictionary definition of climate, includes not only simple averages, such as the climatological January mean surface air temperature, but also, importantly, *statistics that characterize the fluctuations and variations of the climate system*. Variability is of great and even primary interest. Predictions are all about change. Examples of important variations that are aspects of climate include the following:

The seasonal variations of surface air temperature (and many other things)

Systematic day-night temperature differences

The tendency of thunderstorms to occur in late afternoon in many places

The frequency of snow storms

The occurrence, every few years, of "El Niño" conditions, which include unusually warm sea surface temperatures in the eastern tropical Pacific Ocean

The list could easily be extended. Variations like these are important aspects of the climate state. All of them can be described by suitably concocted statistics; for example, we can discuss the average daily minimum and maximum near-surface air temperatures.

It is important to distinguish between "forced" and "free" variations. Forced variations include the day-night and seasonal changes mentioned above, which are externally driven by local changes of solar radiation that are associated with the Earth's rotation on its axis and its orbital motion around the Sun, respectively. Volcanic eruptions can also force climate fluctuations that sometimes last for years. On the other hand, storms and El Niños are examples of unforced or "free" variations that arise naturally through the internal dynamics of the climate system.

For the reasons outlined above, I would modify the dictionary definition to something like this: "Climate is the (in principle, infinite) collection of statistics based on the evolving, geographically distributed state of the atmosphere, including not only simple averages but also measures of variability on a range of time scales from hours to decades."

Since "weather" refers to the state of the atmosphere, the definitions given above make climate appear to be a property of the atmosphere alone. Climate scientists don't think of it that way, though, because any attempt to understand what actually determines the state of the climate, and what causes the climate to change over time, has to take into account the crucial roles played by the ocean (including marine biology), the land surface (including terrestrial biology), and the continental ice sheets. These, together with the atmosphere, make up the four primary components of what is often called the "climate system."

The ocean is about 400 times more massive than the atmosphere and has a heat capacity more than a thousand times larger.[3] When the ocean says "Jump," the atmosphere asks "How high?" Nevertheless, the thin, gaseous atmosphere exerts a powerful influence on the climate. How can the relatively puny atmosphere play such a major role in the much larger climate system?

The explanation has two parts. First, the atmosphere serves as an outer skin, standing between the other components of the climate system and space. As a result, the atmosphere is in a position (so to speak) to regulate the all-important exchanges of energy between the Earth and space, which take the forms of solar radiation coming in and infrared radiation going out.

The second reason is that the atmosphere can transport energy, momentum, and other things from place to place much faster than any other component of the climate system. Typical wind speeds are hundreds or even thousands of times faster than the speeds of ocean currents, which are in turn much faster than the ponderous motions of the continents. The atmosphere (and ocean) can also transport energy through the pressure forces exerted by rapidly propagating fluid-dynamical waves of various kinds.[4]

The climate system is of course governed by the laws of physics. Its behavior can be measured in terms of its physical properties, analyzed in terms of its physical processes, and predicted using physical models. It is influenced by a variety of "external" parameters that are (almost) unaffected by processes at work inside the climate system. These external parameters include the size, composition, and rotation rate of the Earth; the geographical arrangement of oceans, continents, mountain ranges, and so on; the geometry of the Earth's orbit around the Sun; and the amount and spectral distribution of the electromagnetic radiation emitted by the Sun.[5]

Over the past few decades, the possibility of ongoing and future anthropogenic climate change has been widely recognized as a major scientific and societal issue, with huge economic ramifications. As a result, the physical state of the climate system is now being intensely monitored, like the health of a patient with worrisome symptoms. Ongoing changes are being diagnosed. The future evolution of the system is being predicted, using rapidly improving physically based models that run on the fastest computers in the world.

THE COMPOSITION OF THE ATMOSPHERE

The atmosphere is big. Its total mass is about 5×10^{21} g.

The most abundant atmospheric constituents are nitrogen and oxygen. They are very well mixed throughout almost the entire atmosphere, so that their relative concentrations are essentially constant in space and time.

Ozone and water vapor are "minor" but very important atmospheric constituents that are *not* well mixed, because they have strong sources and sinks inside the atmosphere. Ozone makes up less than one millionth of the atmosphere's mass, but that is enough to protect the Earth's life from deadly solar ultraviolet (UV) radiation. Water vapor is only about a quarter of 1% of the atmosphere's mass, but its importance for the Earth's climate, and for the biosphere, would be hard to exaggerate, and it will be discussed at length throughout this book.

The term "dry air" refers to the mixture of atmospheric gases other than water vapor. As shown in Table 1.1, dry air is a mixture of gases. Each gas approximately obeys the ideal gas law, which can be written for a particular gas, denoted by subscript i, as

$$p_i V = N_i k T \qquad (1.1)$$

Here p_i is the partial pressure of the gas, V is the volume under consideration, N_i is the number of particles, k is Boltzman's constant, and T is the temperature. If the gas is in thermal equilibrium, then the temperature will be the same for all gases in the mixture. This is an excellent assumption for dry air, valid up to at least 100 km above the surface. We can write $N_i k = n_i R^*$, where n_i is the number of moles and R^* is the universal gas constant.

Table 1.1

A partial list of the well-mixed gases that make up "dry air," in order of their mass fraction of the atmosphere. Oxygen is present in significant quantities only because of the existence of life on Earth. Three of the six leading constituents are noble gases (argon, neon, and helium). Additional gases (not listed) are present in smaller amounts.

Gas	Molecular form	Molecular mass $g\ mol^{-1}$	Volume fraction ppmv	Mass fraction of the "dry" portion of the atmosphere
Nitrogen	N_2	28.0	781,000	0.755
Oxygen	O_2	32.0	209,000	0.231
Argon	Ar	39.4	9,340	0.0127
Carbon dioxide	CO_2	44.0	390	5.92×10^{-4}
Neon	Ne	20.2	18	1.26×10^{-5}
Helium	He	4.0	5	6.90×10^{-7}
Methane	CH_4	16.0	2	1.10×10^{-7}

The total mass of the gas, M_i, satisfies $n_i = \frac{M_i}{m_i}$, where m_i is the molecular mass. Substituting, we find that

$$p_i = \rho_i \frac{R^*}{m_i} T, \tag{1.2}$$

where $\rho_i \equiv \frac{M_i}{M_i}$ is the density.

The temperature is a measure of the kinetic energy of the random molecular motions. It normally decreases with height in the lower atmosphere, although, as discussed later, it actually increases upward at greater heights. The range of temperatures encountered throughout most of the atmosphere is roughly 200 K to 300 K.[6]

The pressure of a gas is, by definition, the normal component of the force (per unit area) exerted by the moving molecules. In an ideal fluid, the pressure at a point is the same in all directions.

You have probably experienced the increased pressure that the water exerts on your ears (and the rest of your body) at the bottom of a swimming pool. What you are feeling is the weight (per unit horizontal area) of the water above you. At greater depths, there is more water above, it pushes down on you more heavily, and the water pressure increases as a direct result. The density of the water, ρ_{water}, is very nearly constant, so the pressure at a given depth, D, is given by $p = \rho_{\text{water}} gD$, where g is the acceleration of gravity, which is about 9.8 m s^{-2} near the Earth's surface.[7]

Similarly, the air pressure at a given height is very nearly equal to the weight (per unit horizontal area) of the air above. In a "high-pressure" weather system, you are buried under a thicker (more massive) layer of air. With a low-pressure system, the layer of air is thinner. This "hydrostatic" relationship applies to each gas separately because the weights of the gases simply add. For a particular gas, denoted by subscript i, the hydrostatic relationship can be expressed in differential form by

$$\frac{\partial p_i}{\partial z} = -\rho_i g, \tag{1.3}$$

where z is height. The minus sign appears in Equation (1.3) because the pressure increases downward while height increases upward. We use a *partial* derivative of p with respect to z in (1.3) because the pressure also depends on horizontal position and time. The hydrostatic equation, (1.3), expresses a balance between two forces, namely the downward weight of the air and the upward pressure force that arises from the upward decrease of pressure. The balance is not exact. It is an approximation, which is another way of saying that it has an error because something has been neglected. That something is the actual acceleration of the air in the vertical direction. Further discussion is given in Chapter 3.

By combining the ideal gas law with the hydrostatic equation, we find that

$$\frac{1}{p_i}\frac{\partial p_i}{\partial z} = -\frac{m_i g}{R^* T}. \tag{1.4}$$

As you probably know, g decreases upward in proportion to the square of the distance from the center of the Earth. The Earth's atmosphere is very thin, though, so the variations of g with height inside the atmosphere are negligible for most purposes, and we will neglect them here. Suppose that the temperature varies slowly with height. Treating it as a constant,[8] we can integrate both sides of (1.4) to obtain

$$p_i(z) = \left(p_i\right)_S e^{-\frac{z}{H_i}}, \tag{1.5}$$

where the surface height is taken to be zero, $\left(p_i\right)_S$ is the surface partial pressure of gas i, and

$$H_i \equiv \frac{R^* T}{m_i g} \tag{1.6}$$

is called the "scale height" of the gas. Equation (1.5) says that the partial pressure decreases upward exponentially away from the surface, at a rate that depends on the molecular mass of the gas and also on the temperature, which, again, is the same for all of the gases in the mixture. By putting in some numbers, you will find that a typical value of the scale height of nitrogen in the Earth's atmosphere is about 8,000 m. According to Equation (1.5), the partial pressure of nitrogen decreases upward by a factor of e in 8,000 m.

How heavy is a column of air? The total pressure (due to the weight of all atmospheric constituents) is typically about 100,000 Pa near sea level, where a Pa (pascal) is defined to be a newton (Nt) per square meter. For comparison, the weight of a typical car is about 20,000 Nt, so the weight of an air column per square meter is roughly equivalent to the weight of five cars piled on top of each other, over one square meter of a junk yard. That's pretty heavy. The total pressure at an altitude of 12 km is about 20,000 Pa, roughly 5 times less than the surface pressure.

By pushing the ideas presented above just a bit further, you should be able to show that the partial density of a gas also decreases upward exponentially, following a formula very similar to (1.4). The total density of the air near sea level is typically about 1.2 kg m^{-3}. The total density at an altitude of 12 km is about 5 times less. As an example, the density of atmospheric oxygen decreases exponentially upward, which is why hiking is more challenging at higher altitudes.

As shown in Table 1.1, the various atmospheric gases have different molecular masses, and as a result they have different scale heights. This suggests that the relative concentrations of the various gases should vary with height, so that the heaviest species would be more concentrated near the ground and the lighter species more concentrated higher in the atmospheric column. The process that could lead to such a result is called "diffusive separation" because it depends on the microscopic shaking due to molecular motions, in the presence of gravity. Diffusive separation is a real process, but it is negligible in the lower atmosphere because the turbulent winds act like a powerful mixer that homogenizes the blend of gases. The result is that, except for ozone, the concentrations (*not* the densities) of the gases that make up dry air, for example, nitrogen, oxygen, argon, and carbon dioxide, are observed to be nearly homogeneous, both horizontally and vertically, up to an altitude of about 100 km. Above that level, diffusive separation does become noticeable.

As an example, Figure 1.1 shows the variations of density, pressure, and temperature with height, from the surface to an altitude of 50 km, based on the U.S. Standard Atmosphere (http://modelweb.gsfc.nasa.gov/atmos/us_standard.html).[9]

Vertical profiles are called "soundings," a term borrowed from oceanography. The approximately exponential upward decreases of density and pressure are obvious. The temperature decreases upward for the first 10 km or so, remains almost constant for the next 10 km,

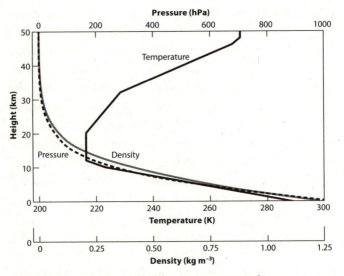

Figure 1.1. Typical variations of density, pressure, and temperature with height, from the surface to an altitude of 50 km, which is in the upper stratosphere.

The exponential upward decreases of density and pressure are clearly evident. The temperature variations are quite different; near the surface it decreases upward, but above about the 20 km level it actually increases upward, for reasons discussed in the text. The scales for density and pressure start at zero, while the scale for temperature starts at 200 K. This figure is based on the U.S. Standard Atmosphere.

and then begins to increase upward. This vertical distribution of temperature will be explained later.

Water vapor is an important exception to the rule that the atmosphere is well mixed. There are two main reasons for this. First, water vapor enters the atmosphere mostly by evaporation from the ocean, so it has a tendency to be concentrated near the surface. More importantly, water vapor can condense to form liquid or ice, which then falls out of the atmosphere as rain or snow. Condensation happens when the actual concentration of vapor exceeds a saturation value, which depends on temperature. In effect, the water vapor concentration is limited to be less than or equal to the saturation value. At the colder temperatures above the surface, saturation occurs more easily, and the water concentration decreases accordingly. The fact that water can change its phase in our atmosphere is critically important for the Earth's climate. Much additional discussion is given in later chapters.

In a warm place near the surface, water vapor can be as much as 2% of the air by mass. You would drip with sweat in a place like that. In a cold place, the air might contain 0.1% water vapor by mass; your skin would tend to dry out and crack under those conditions. In the stratosphere, the mass fraction of water vapor is typically just a few parts per million.[10]

In addition to the various gases discussed above, each cubic centimeter of air contains many, sometimes hundreds, of tiny liquid and solid particles called aerosols. Aerosols include dust, much of which comes from deserts, where it is scoured away from the land surface by the drag force associated with the winds. The dust can be carried over vast distances. For example, Asian dust is carried eastward across the Pacific to California, and Saharan dust is carried westward across the Atlantic to Florida and eastern South America. Another important type of aerosol is sea salt. The spray whipped up by winds near the sea surface evaporates, leaving behind tiny salt particles, suspended in the air, that can then be carried long distances by the winds. Pollutants emitted by human activities are another major source of aerosols. These anthropogenic aerosols include sulfur compounds and also carbon compounds such as soot.

The most obvious liquid and solid particles in the atmosphere are the liquid water drops and ice crystals that make up clouds. As a matter of terminology, however, atmospheric scientists conventionally reserve the term "aerosol" to refer to the wide variety of *noncloud* liquid and solid particles in the air. Cloud particles range in size from about 10 microns (a micron is 10^{-6} meter) to a centimeter or so for large rain drops and snowflakes. Particles smaller than about 0.1 mm fall slowly because their motion is strongly limited by drag, so to a first approximation they can be considered to move with the air like a gas. Particles that fall more quickly are said to "precipitate"; their fall speeds are close to the "terminal velocity" at which their weight is balanced by aerodynamic drag. The drag increases with the density of the air, and with the square of the fall speed. Large rain drops fall at about 5 m s^{-1}, relative to the air. If such drops find themselves in an updraft with a speed faster than 5 m s^{-1}, they will actually be carried upward by the air.

The liquid and ice particles that make up clouds are typically "nucleated" on aerosols; the variable abundance of these cloud-nucleating particles, which are called cloud condensation nuclei (CCN), is a factor influencing the formation, number, and size of cloud particles. The number of CCN does not strongly influence the probability of cloud formation or the area-averaged rate of precipitation, under realistic conditions. The fascinating complexities of clouds are discussed further in later chapters.

Clouds and aerosols make up a tiny fraction of the atmosphere's mass, but they powerfully affect the flow of electromagnetic radiation through the atmosphere. This is also discussed further in later chapters.

LAYERS

There is no well-defined "top of the atmosphere" because the density of the air just continues to decrease exponentially toward zero at great heights. Nevertheless, all but about one millionth of the mass of the atmosphere lies below a height of 100 km, so it is reasonable (and somewhat conventional) to use this height as the atmosphere's top. Although 100 km sounds like a pretty thick layer of air, the radius of the Earth is a much larger 6,400 km. Relative to the Earth as a whole, the atmosphere is almost like a coat of paint, as can be seen in Figure 1.2. Life on Earth critically depends on that thin, fragile, mostly transparent, wispy shell.

The atmosphere is conventionally divided into layers. The lowest layer is the *troposphere*, which is in direct contact with the *lower boundary*, a term used by atmospheric scientists to refer to the top of the ocean and land surface. As discussed later, the lower boundary can be viewed as an energy source for the atmosphere. Roughly speaking, the atmosphere is "heated" by contact with the boundary;

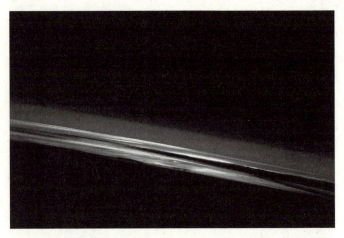

Figure 1.2. A photograph of the atmosphere as seen from space; the Sun is just below the horizon.

If the figure were in color, you would see that the lower atmosphere looks red because the blue photons emitted by the Sun are scattered away from the line of sight. A few clouds can be seen as dark blobs, blocking the Sun's rays.

Source: http://upload.wikimedia.org/wikipedia/commons/1/13/sunset_from_the_ISS.JPG

the heat enters directly into the base of the troposphere. In response, the troposphere churns like a pot of water on a stove, as buoyant chunks of air break away from the lower boundary and float upward, carrying energy (and other things) with them. The upper-level air is cooled by emitting infrared radiation to space. Once cooled, it sinks back to near the surface, where it is heated again, and the cycle repeats, in an irregular and chaotic fashion. The buoyancy-driven or "convective" motions act like a gigantic heat pump, cooling near the surface and warming above.

You may be wondering, if buoyancy pushes warm air up and cold air down, then shouldn't we find the warm air on top and the cold air below? Why does the temperature typically decrease upward? A partial explanation is that sinking air is warmed by compression as it moves to greater pressures near the surface and rising air is cooled by expansion as it moves to smaller pressures aloft. Further discussion is given in Chapter 3.

The small- and large-scale convective motions that are driven by surface heating and upper-level cooling extend through the lowest 10 or even 20 km of the atmosphere, depending on place and season. This layer of active weather is the troposphere. It contains about 80% of the atmosphere's mass.

The lowest portion of the troposphere is often called the *boundary layer* because it is strongly affected by direct contact with the lower boundary, that is, the ocean or the land surface. The boundary layer is turbulent; in fact, this is a key part of its definition.

The top of the troposphere is called the "tropopause." You have probably been there. Commercial airliners typically cruise near the tropopause level, at a height of about 12 km and a pressure of about 20,000 Pa.

You may feel a bit surprised that the troposphere has a well-defined top. As discussed in a later chapter, active weather, which is a characteristic property of the troposphere, is a kind of turbulence. Turbulent masses of air often have well-defined boundaries because turbulence tends to grow by active annexation of the surrounding, quiet fluid, as if by an advancing army.

Above the tropopause is the stratosphere. It is meteorologically quiet, compared to the troposphere. A distinctive property of the stratosphere is that the temperature generally increases upward there, as can be seen in Figure 1.1. The upward temperature increase is caused by heating of the middle and upper stratosphere due to the absorption of UV radiation from the Sun. The UV is actually absorbed by ozone (O_3), which is thereby converted into molecular oxygen (O_2) and atomic oxygen (O). The O_2 and O then recombine (in the presence of other species) to create another ozone molecule, so that there is no net loss of ozone. The net effect is a continual destruction and regeneration of ozone, accompanied by a heating of the air. The ozone cycle is an essential characteristic of the stratosphere.

The absorption of solar UV by ozone is important for life because UV causes biological damage if and when it reaches the biosphere. The actual amount of ozone in the atmosphere is tiny. Like water vapor, ozone is an atmospheric constituent that is *not* well mixed. In the middle stratosphere, about 30 km above the ground, the ozone concentration reaches a maximum, but it is a very small maximum: only about 1/100,000 of the air, depending on place and season. In the troposphere, ozone is much less abundant, except where it is artificially introduced as a pollutant.

The top of the stratosphere is called the stratopause. It is typically about 60 km above the surface. The troposphere and stratosphere together contain about 99.99% of the mass of the atmosphere.

....

It is conventional to define more layers above the stratopause, including the following:

The *mesosphere*, which extends from the stratopause to about 90 km, and within which the temperature decreases upward. The stratosphere and mesosphere together are often referred to as "the middle atmosphere"—no relation to Tolkien's Middle Earth. Reentering spacecraft heat up in the mesosphere, and most meteors burn up there. The mesopause coincides with what is sometimes called the "turbopause," the level above which mixing by the winds is no longer dominant.

The *thermosphere*, from the mesopause to about 500 km. Here the atmospheric gases are no longer well mixed and fractionate by their molecular weights. Molecular viscosity is strong enough to suppress turbulent mixing. The air is ionized by UV radiation, and as a result its motions can be influenced by electromagnetic fields. The temperature increases upward and is sensitive to the level of solar activity. The International Space Station orbits within the thermosphere.

The *exosphere*, from 500 km to an indefinite height. Within the exosphere, the atmosphere consists of rarely colliding individual molecules, mostly hydrogen and helium. Some of the molecules attain escape velocity and leave the Earth behind.

In this book, we will focus mainly on the troposphere.

THE WINDS

The wind is the motion of the air. Typical wind speeds range from a few meters per second or less near the surface, to 50 m s^{-1} in the jet streams near the tropopause level, to 100 m s^{-1} in a strong tornado. These air currents arise in response to forces acting on the air, including pressure forces, which are often produced by spatially varying patterns of heating and cooling. We say that the circulation of the atmosphere is "thermally driven."

The term "scale" has already been used several times in this book. Atmospheric processes often have characteristic space and/or time scales. The concept of scale is particularly useful in connection with the winds and ocean currents. Spatial and temporal scales tend to increase or decrease together. Turbulent eddies are meters to hundreds of meters across and last for seconds to minutes. Thunderstorms are a few kilometers across and last for an hour or two. Winter storms can be thousands of kilometers across and last for several days. Monsoons span thousands of kilometers and persist for months. These diverse scales of motion can strongly interact with each other. Large-scale weather systems can excite small-scale turbulence. The turbulence, in turn, exerts drag forces and other influences on the large-scale weather. Such "scale interactions" are extremely important and will come up repeatedly throughout this book.

Because the atmosphere is a thin shell on the large spherical Earth, air can move much further horizontally than it can vertically. As a result, the large-scale motion of the air is mostly horizontal, and typical horizontal wind speeds are generally hundreds or even thousands of times faster than typical vertical wind speeds. On smaller scales, however, the vertical winds are sometimes just as fast as the horizontal winds; this can be the case in thunderstorms, for example, which can contain narrow but intense updrafts and downdrafts with speeds of tens of meters per second—not a friendly environment for aerial navigation. The winds carry energy, moisture, and other things, from place to place. Even moderate wind speeds are fast enough to carry air between any two points on the Earth in a few weeks.

The atmosphere is a closed system; air moves around, but, with minor exceptions, it does not enter or leave the atmosphere. It is therefore natural to speak of atmospheric "circulations." Some of these circulations are closely associated with familiar weather phenomena and reoccur, with minor variations, often enough so that we find it useful to give them names. Examples are thunderstorms, fronts, jet streams, tropical cyclones, and monsoons.

One of the most important factors influencing the circulation is the rotation of the Earth, which has profound effects on large-scale circulations. Because of the Earth's rotation, a point on the Earth's surface is moving toward the east at a speed that depends on latitude and is fastest at the Equator. Here is a number that is easy to remember: At the Equator, a point on the Earth's surface is moving toward the East at about 1,000 miles per hour (about 450 m s^{-1}). The Earth's rotation promotes the formation of beautiful atmospheric vortices. The vortices of many sizes and varieties are not details; they are essential features of the winds. Most of the kinetic energy of the atmosphere is associated with vortices of one kind or another.

Mountain ranges exert a major influence on the circulation because the moving air must go over or around such obstacles. The major mountain ranges of the world cause prominent wavy patterns in the global circulation. Some of the waves span thousands of kilometers and strongly influence the weather. Mountains also influence the pattern of atmospheric heating and cooling.

Small-scale motions are often driven by buoyancy forces associated with clouds; these are the convective circulations mentioned earlier. Buoyancy-driven circulations occur on a wide range of spatial scales, from tens of meters to thousands of kilometers. The circulating air moves not only vertically but also horizontally, along paths that sometimes approximate closed loops. These circulations are the wind fields that are associated with what we call weather. Familiar examples of convective circulations include thunderstorms, which are comparable in width to a city; continent-sized winter storms in middle latitudes; and monsoons, which can be comparable in horizontal scale to the radius of the Earth.

LATENT HEAT

From a human perspective, precipitation is perhaps the single most important climate variable. Less obviously, cloud and precipitation processes are leading actors in the physical climate system and strongly influence the global circulation of the atmosphere. One reason for this is that water has an enormous latent heat of vaporization, which can be defined as the energy required to evaporate a given mass of liquid. The latent heat of water is about 2.5 MJ kg^{-1}. Here is an amazing fact: the 2.5 MJ needed to evaporate one kilogram of water is equivalent to the kinetic energy of one kilogram of mass (which could be the same kilogram of water) moving at about 2,200 meters per second, or 7 times the speed of sound! For comparison, the latent heat of nitrogen, the most abundant atmospheric constituent, is only about 0.20 MJ kg^{-1}, less than one-twelfth that of water.[11]

When you climb out of a swimming pool, water evaporates from your skin. The energy needed to evaporate the water comes from your skin, and as a result the evaporation makes you feel cold. That same 2.5 MJ kg^{-1} of latent heat is "released" when condensation occurs. You may have experienced latent heat release in a sauna. Liquid water that is tossed onto the hot coals of a sauna flashes into the air as vapor and then condenses onto your relatively cool skin. The latent heat release that accompanies this condensation makes your skin feel hot.

The gigantic latent heat of water is one of the reasons that "moist processes" (atmospheric science jargon for processes involving phase changes of water) are very important for the Earth's climate. This will be discussed in a later chapter.

A FIRST LOOK AT THE ENERGY CYCLE

Good advice for students of the climate system is "Follow the energy." That is the overarching theme of the next three chapters of this book. A broad overview of the energy flow is given in Figure 1.3. The climate system absorbs energy in the form of sunlight, mainly in the tropics and mostly at the Earth's surface. Clouds, ice, and bright soils modulate this solar absorption by reflecting some of the sunlight back to space. A substantial portion of the absorbed solar energy is used to evaporate water from the

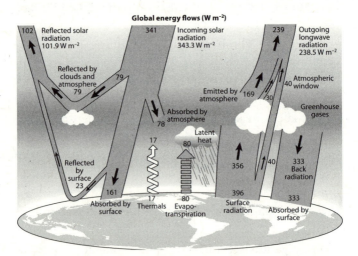

Figure 1.3. An overview of the flow of energy in the climate system. The global annual mean Earth's energy budget for the March 2000 to May 2004 period (W m^{-2}). The broad arrows indicate the schematic flow of energy in proportion to their importance.

Source: Based on a figure in Trenberth et al. (2009).

tropical oceans. A larger portion enters the atmosphere in the form of infrared radiation emitted by the surface. The gases and clouds of the atmosphere absorb some of this, and reradiate the energy, again as infrared, both downward to the surface and upward to space.

Overall, considering both solar and infrared radiation, the atmosphere is radiatively cooled. The radiative cooling is balanced by the latent heat released when the water evaporated from the ocean recondenses to form clouds. In this and other ways, the Earth's energy and water cycles are closely linked. Chapters 3, 5, and 6 explore these links in more detail.

Atmospheric processes convert a small portion of the thermodynamic energy into the kinetic energy of atmospheric motion. The winds and the ocean currents carry the energy around from place to place on the Earth, cooling some regions and warming others, something like the heating, ventilating, and air conditioning system in a building.

The air also has gravitational potential energy, usually shortened to "potential energy," which increases (per unit mass) with height above the surface. The kinetic and potential energies combined are sometimes called the mechanical energy.

Ultimately, the energy that was originally provided by the Sun flows back out to space in the form of infrared radiation, called the "outgoing longwave radiation," abbreviated as OLR. The atmosphere strongly absorbs the infrared radiation emitted by the lower boundary and reemits the energy at the colder temperatures found aloft. Measurements show that, on time scales of a year or more, the globally averaged infrared emission by the Earth balances the globally averaged solar absorption very closely, within a couple tenths of a percent. The imbalance is believed to be just at the limit of what can currently be measured.

Radiative processes are key at both the beginning and the end of the energy story. They deserve a chapter of their own.

THE EARTH'S RADIATION BUDGET

ELECTROMAGNETIC RADIATION IS (ALMOST) THE ONLY way that the Earth can exchange energy with the rest of the universe.[1] The radiation can be divided into categories, by wavelength:

UV radiation, roughly from 0.1 to 0.4 μm, which is emitted by the Sun, absorbed in the stratosphere where it promotes the creation of ozone, and damaging to living things when it reaches the Earth's surface

Visible light, roughly from 0.4 to 0.8 μm, which contains most of the Sun's energy output and is important for photosynthesis in plants and for human sight

Near-infrared radiation, roughly from 0.8 to 4 μm, which is emitted by the Sun

Thermal infrared radiation, roughly from 4 to 50 μm, which is emitted by the Earth and is the Earth's primary way of returning energy to space

Other wavelengths (e.g., X-rays and radio waves) are only very weakly emitted by the Sun (and the Earth) and are not important for the Earth's climate. As discussed at length below, averaged over the Earth and averaged over a year, the solar radiation absorbed is very nearly equal to the infrared radiation emitted back to space. The Earth is close to energy balance.

SOLAR RADIATION

The Sun provides an intense supply of electromagnetic radiation from a small, point-like patch of sky, that is, a small solid angle. The incoming solar radiation at the top of the atmosphere is called the "insolation," or the "solar irradiance." It is also called the "solar constant," and in fact the Sun is a remarkably but not perfectly constant energy source. Calibrated measurements from satellites, which have been available since about 1980, show that, during the past 30 years, the total rate of energy output by the Sun has varied by only about ±0.1% or about ±1 W m^{-2}. Fortunately for us, the Sun is Old Reliable.

The total (summed over all wavelengths) solar irradiance, at the mean radius of the Earth's not-quite-circular orbit, is about 1,365 W, which is roughly the energy used by a hair dryer, per m^{-2}. Another way to get an intuitive grasp of this number is to imagine fourteen 100 W light bulbs per square meter.[2] Another is to consider that 1,365 W m^{-2} is the same as 1.365 GW km^{-2}, that is, equivalent to the energy output of a large power plant for each square kilometer of the Earth's "projected" circular area normal to the solar beam, that is, πa^2, where a is the radius of the Earth.

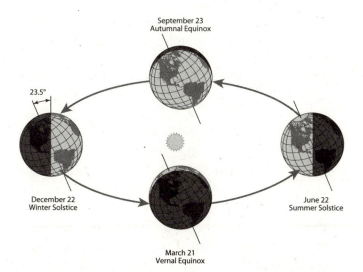

Figure 2.1. Changes in the Sun-Earth geometry as the Earth moves in its orbit.

The Earth's axis is tilted with respect to the plane of its orbit. As the tilted Earth revolves around the Sun, changes in the distribution of sunlight cause the succession of seasons.

Source: http://oceanservice.noaa.gov/education/yos/resource/ JetStream/global/global_intro.htm.

To an observer riding on the Earth, the Sun appears to move in the sky as the Earth spins on its axis and travels through its orbit. The seasonal cycle of solar radiation at a particular location on the Earth is determined by the geometry of the Earth's orbit (see Figure 2.1), including

The obliquity, which is the angle that the Earth's axis of rotation makes with the normal to the orbital plane

The eccentricity, which is a measure of the departure
of the shape of the orbit from a perfect circle

The dates of the equinoxes (or, alternatively, the dates
of the solstices), which matter because the eccentricity is not zero

All of these vary over geologic time due to the gravitational influences of the other planets of the Solar System (e.g., Crowley and North, 1991). In middle and high
latitudes (in other words, away from the Equator), the
seasonal variation of the insolation is mostly due to the
obliquity. For example, the Northern Hemisphere receives
more sunshine in the northern summer, when it is tilted
toward the Sun, and less in the northern winter, when it is
tilted away from the Sun. The Earth's obliquity varies only
slightly, even on geologic time scales, because the gravitational influence of the Earth's large moon is a stabilizing
influence (Laskar et al., 1993). In contrast, Mars, with its
two tiny moons, undergoes drastic obliquity changes on
time scales of hundreds of thousands of years.

In the present era, the Earth is slightly closer to the
Sun in January and slightly farther away in July. As a result, the Earth receives about 6% more energy from the
Sun in January than July. Over thousands of years, the
month in which the Earth is closest to the sun gradually
changes. This "precession of the equinoxes" is important
for the ice ages because strong summer insolation in the
high latitudes of the Northern Hemisphere ensures summer melting of the snow and so prevents the formation of
ice sheets on North America and Europe. The Northern

Hemisphere is key because it has much more land surface (much less ocean) than the Southern Hemisphere, and it is the land surface that can support ice sheets.

The total solar energy *incident* on the Earth is $S\pi a^2$, where $S \cong 1365$ W m^{-2} is the energy per unit area per unit time in the "solar constant," and a is the radius of the Earth. Here πa^2 is the circular cross-sectional area that that the spherical Earth presents to the solar beam. The total solar energy *absorbed* by the Earth, divided by the area of the Earth's spherical surface, can be written as

$$S_{abs} = S\left(\frac{\pi a^2}{4\pi a^2}\right)(1-\alpha)$$
$$= \frac{1}{4}S(1-\alpha) \cong 240 \text{ W m}^{-2} \quad \text{(annual mean).}$$

(2.1)

Here S_{abs} is the average absorbed solar energy per unit area and per unit time and α is the *planetary albedo*, defined as the globally averaged energy scattered back to space divided by the globally averaged energy coming in from the Sun. The reflected solar radiation, per unit area, is $S\alpha$. The Earth's albedo is close to 0.3, independent of season; this number has been accurately known only since the advent of satellite data in the 1970s. In Equation (2.1), the rate of energy absorption is referred to the total surface area of the Earth, that is, $4\pi a^2$, rather than to the cross-sectional or "projected" area presented to the solar beam, which is four times smaller. The motivation is to include both the day and night sides of the Earth in the global average.

As you know from your own experience, the insolation at a given location varies both diurnally (i.e., between day and night) and seasonally. Another way of thinking about the day-night cycle is that, at a given moment, the insolation varies strongly with longitude. Because a year is much longer than a day, the daily mean insolation is (almost) independent of longitude, but it varies strongly with latitude in a way that depends on the season, as summarized in Figure 2.2. The data shown in the figure have been "zonally averaged." This is jargon that we will use a lot, so it is important to remember what it means: A zonal average is an average over all longitudes at a given latitude. In other words, it is an average around a latitude circle.

The shapes of the contours in Figure 2.2 need some explanation. As we move from the Equator toward the summer pole, there are two competing effects. The sun moves lower in the sky, but the day becomes longer. As Figure 2.2 shows, the first effect dominates at low latitudes, so that insolation initially decreases with increasing latitude. At higher latitudes, the length-of-day effect becomes dominant, so the insolation increases poleward (in the summer), from a minimum in the midlatitudes. Around the time of the solstices, no insolation at all occurs near the winter pole ("polar night"), but at the same time, near the summer pole, the daily mean insolation is very strong despite low sun angles, simply because the sun never sets ("polar day").

The solar irradiation outside the atmosphere is closely approximated by the emission from a blackbody with a temperature of 5,900 K.[3] It is almost entirely

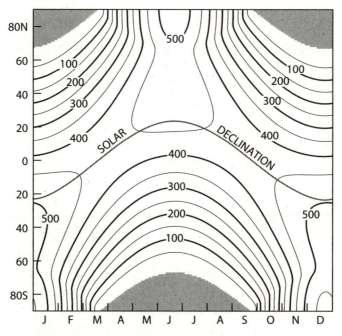

Figure 2.2. The seasonal variation of the zonally (or diurnally) averaged insolation at the top of the atmosphere.

The units are W m^{-2}.

contained between the wavelengths of 0.2 μm (UV) and 3 μm (near infrared). The solar irradiation at sea level is diminished because of absorption and scattering by the atmosphere, including gases, clouds, and aerosols. The pattern of absorption of solar radiation over the Earth is modulated by the distributions of land and sea, and the time-varying distributions of snow and ice, and clouds and aerosols.

TERRESTRIAL RADIATION

The infrared radiation emitted by the solid Earth, oceans, and atmosphere is called "terrestrial radiation," or sometimes "thermal infrared." It is also called "longwave radiation" because of its relatively long wavelengths. The gases, clouds, and aerosols that make up the atmosphere absorb and emit (and to a lesser extent scatter) longwave radiation, for wavelengths between about 3 µm and 70 µm. The atmospheric emitters and absorbers include water vapor, clouds, carbon dioxide, methane, nitrous oxide, ozone, and a number of other trace gases. The gases that absorb and emit longwave radiation are called greenhouse gases. Notice that the gaseous absorbers/emitters listed here all have three or more atoms per molecule (H_2O, CO_2, CH_4, N_2O, and O_3); molecules with only two atoms, such as molecular nitrogen and oxygen, do not absorb or emit infrared radiation. Naturally, the strength of absorption and emission depends on the concentrations of the relevant gases; all except carbon dioxide are highly variable in space and time.

For a given concentration of an emitter, the rate of infrared emission increases with temperature. Air that is up high can efficiently emit to space because there is little air above to block the infrared photons on their way out of the atmosphere. On the other hand, air that is up high is cold, so emission is weak for a given concentration of the emitter.

As mentioned in Chapter 1, the terrestrial radiation flowing out through the top of the atmosphere is called the outgoing longwave radiation, or the OLR for short. The globally averaged OLR varies only slightly with season, and from year to year. Its average value is a bit less than 240 W m^{-2}, which is the flux that would be emitted by a blackbody at a temperature of about 255 K, far colder than the freezing point of water.[4] The "effective altitude" for infrared emission by the Earth-atmosphere system is near 5 km above sea level because that is the level where the average temperature is about 255 K. This simply means that the OLR is equivalent to that from a blackbody whose temperature is that of the atmosphere near the 5 km level. Roughly speaking, then, atmospheric motions must carry energy upward from the surface through the first 5 km of the atmosphere and infrared emission carries the energy the rest of the way out to space. Chapters 3 and 4 explain how this works.

The Earth's brightness temperature is the temperature of a blackbody that emits the same infrared radiation per unit area as the Earth. A body with a warm brightness temperature emits strongly. The difference between the Earth's brightness temperature and its actual surface temperature is a measure of the infrared opacity of the atmosphere. A partially opaque atmosphere "hides" the surface temperature from an observer in space.

ABSORPTION AND SCATTERING

Figure 2.3 summarizes the absorption and scatting of both solar and terrestrial radiation by the gaseous constituents

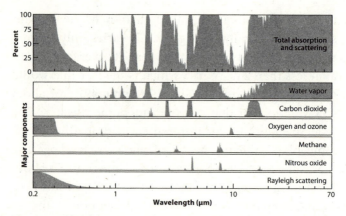

Figure 2.3. The absorption and scattering spectra for major gases in the Earth's atmosphere.

For both panels of the figure, the values along the horizontal axis are the wavelengths of the radiation, with a logarithmic scale. Recall that visible light has wavelengths in the range 0.4 to 0.8 μm, so it is concentrated on the left-hand side of the axis. In the upper panel, the vertical axis shows the percentage of a particular wavelength that is absorbed or scattered back to space by atmospheric gases. The lower panel shows the contributions to absorption and scattering from various constituents, namely water vapor, carbon dioxide, oxygen and ozone, methane, nitrous oxide, and other gases (mainly nitrogen). Rayleigh scattering is the scattering of radiation by gases; it is distinguished from scattering by clouds and aerosols.

Source: http://www.globalwarmingart.com/images/4/4e/ Atmospheric_Absorption_Bands.png.

of the Earth's atmosphere, and also the effects of "Rayleigh scattering," which occurs when photons ricochet off of molecules of nitrogen and other gases.

The absorption of solar radiation, between 0.2 μm and 3 μm (roughly the left half of the diagram), is mostly due to ozone, water vapor, and CO_2. Most of Figure 2.3 is devoted to infrared wavelengths, longer than 0.8 μm. Water vapor absorbs in several fairly discrete infrared bands and even a little bit in the visible part of the spectrum. CO_2 absorbs strongly between 2 and 3 μm, where it overlaps with water vapor absorption, and also between 4 and 5 μm, where water vapor is nearly transparent. As a result, some radiation that is not absorbed by water vapor can be absorbed by CO_2. The two gases together block terrestrial radiation to space more completely than either alone.

THE FLOW OF RADIATION

Let's think about the flow of radiation through the atmosphere, from space all the way down to the Earth's surface, and back out to space.

We'll start with the solar radiation coming in at the top. Although the atmosphere does absorb and scatter (or "reflect") back to space a substantial portion of the incoming solar radiation, most of the solar energy penetrates through to the Earth's surface, especially in the visible part of the spectrum, which includes wavelengths between about 0.4 μm and 0.8 μm. Of the 240 W m^{-2} of solar radiation that is absorbed by the Earth-atmosphere system, about two-thirds, or about 160 W m^{-2}, is absorbed by the Earth's surface. Only about $240 - 160 = 80$ W m^{-2} of solar radiation is absorbed by the atmosphere, roughly one-third of all the solar radiation absorbed by the Earth-atmosphere system. To a first approximation, the atmosphere is transparent, especially to visible radiation. That presumably explains why our eyes are sensitive to the wavelengths that we call visible light.

The surface reflects, or scatters, a fraction of the solar radiation impinging on it. The fraction reflected is called the surface albedo. It depends on surface composition and sun angle, among other things. The ocean has an albedo close to 0.06 when the sun is high in the sky; that is, it is quite dark. At low sun angles, however, the ocean can reflect considerably more; think of "sun glint." The ocean's albedo is much larger (~0.8) when sea ice forms, especially when the ice is covered by snow. The albedo of the land surface is more complicated. It varies widely due to differing compositions of the soil or rock at the surface, differing types and amounts of vegetation cover, and the presence or absence of snow. The surface albedo has a strong effect on the planetary albedo defined earlier, but the latter is also influenced by cloudiness.

The surface radiation budget is summarized in Table 2.1. The surface experiences a radiative heating, partly

Table 2.1
Components of the globally and annually averaged surface radiation budget. A positive sign means that the surface is warmed.

Absorbed solar (SW)	161 W m^{-2}
Downward infrared (LW↓)	333 W m^{-2}
Upward infrared (LW↑)	−396 W m^{-2}
Net longwave (LW)	−63 W m^{-2}
Net radiation (SW + LW)	98 W m^{-2}

due to the solar radiation that it absorbs, denoted by SW, *and also due to the even larger downward infrared emitted by the atmosphere*, denoted by $(LW\downarrow)_S$. The sum of the two is 494 W m^{-2}, which is much larger than the insolation at the top of the atmosphere and approximately twice as large as the OLR.

Naturally, the surface emits infrared radiation back upward into the atmosphere. The ocean and land surfaces both behave approximately like blackbodies, so that the upward infrared satisfies

$$(LW\uparrow)_S \cong \sigma_{SB}T_S^4. \qquad (2.2)$$

Here $(LW\uparrow)_S$ is the infrared radiation emitted by the surface, and $\sigma_{SB} \cong 5.67 \times 10^{-8}$ W m^{-2} K^{-4} is the Stefan-Boltzmann constant. The globally averaged surface temperature is about 288 K, so, putting numbers into (2.2), we find that $(LW\uparrow)_S$ is about 390 W m^{-2}. Because $(LW\uparrow)_S$ is proportional to the fourth power of the surface temperature, it increases very rapidly as the surface warms.

Much less infrared energy leaves the atmosphere to space (240 W m^{-2}) than enters the atmosphere from below (390 W m^{-2}). For the real Earth, the OLR is only about 62% as large as the upward emission of infrared by the Earth's surface (240/390 = 0.62). Writing

$$OLR = \varepsilon_B \sigma_{SB} T_S^4, \qquad (2.3)$$

we can say that the *bulk emissivity of the Earth*, denoted by ε_B, is 0.62. We can also write $OLR = \sigma_{SB}T_{\text{brightness}}^4$, where $T_{\text{brightness}} = 255$ K is the Earth's brightness temperature. It follows that $\varepsilon_B = \frac{T_{\text{brightness}}^4}{T_s^4}$. The bulk emissivity will come up again later.

If the Earth had no atmosphere, ε_B would be equal to one. The departure of ε_B from unity, that is, $1 - \varepsilon_B$, is a measure of the opacity of the atmosphere to infrared radiation. The actual value of ε_B is influenced by several factors, including the composition of the atmosphere (gases, clouds, and aerosols) and the temperature sounding (i.e., the vertical profile of atmospheric temperature).

The Earth's relatively low bulk emissivity means that the atmosphere strongly absorbs the infrared emitted by the surface. The same gases and clouds that absorb infrared radiation also emit it, both upward toward space and downward toward the Earth's surface. The downward emission tends to cool the atmosphere and warm the surface. It is a very important and perhaps counterintuitive fact that *the Earth's surface actually absorbs more energy from downwelling infrared, emitted by the atmosphere (333 W m^{-2}), than it gains from the sun (161 W m^{-2})!* The downwelling infrared at the surface is part of what is meant by "the greenhouse effect." The difference between the surface emission and the OLR is another measure of the greenhouse effect.[5]

How does infrared radiation make its way upward through the atmosphere? Consider a spot on the Earth's surface. The spot emits infrared radiation, according to its temperature, which affects both the rate of emission (energy per unit area per unit time) and the range of infrared wavelengths that is emitted. Imagine that you are a thermal-infrared photon, emitted from the spot. Will you

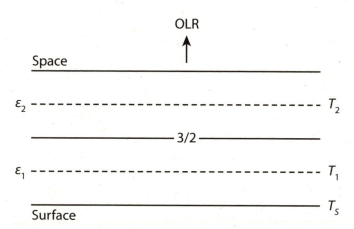

Figure 2.4. An idealized model in which the atmosphere is represented by just two layers, with different temperatures and emissivities.

The lower layer is denoted by subscript 1, and the upper layer by subscript 2. The interface between the two layers is denoted by subscript 3/2.

make it out to space? There is an obstacle course in your way, consisting of absorbing gases, clouds, and aerosols.

To see how this works, we now analyze a very simple and highly idealized example. We represent the atmosphere with just two layers,[6] as shown in Figure 2.4.

Each layer has a temperature and an emissivity and emits infrared at the rate $\varepsilon_i \sigma_{SB} T_i^4$, where the subscript i can be either 1 or 2, denoting the lower or upper atmospheric layer. The emissivities of the two layers are assumed to be known. We also include the Earth's surface, assumed to be a blackbody (emissivity one), with temperature T_S, emitting infrared radiation upward at the rate $\left(LW \uparrow \right)_S = \sigma_{SB} T_S^4$.

We start by figuring out the expressions for $LW \uparrow$ at each layer edge, starting at the Earth's surface, where $\left(LW \uparrow\right)_S = \sigma_{SB} T_S^4$. At the interface between the two atmospheric layers, denoted by subscript 3/2, the upward infrared energy flux is given by

$$\left(LW \uparrow\right)_{3/2} = \left(1 - \varepsilon_1\right)\sigma_{SB} T_S^4 + \varepsilon_1 \sigma_{SB} T_1^4. \tag{2.4}$$

The first term represents the "surviving" part of the upward surface emission, that is, the part that is not absorbed as it passes through layer 1. A fraction ε_1 of the surface emission has been absorbed by layer 1 and therefore does not make it to the interface between the layers. The second term on the right-hand side of (2.4) represents the upward emission from layer 1; the same amount is emitted downward, and this is accounted for below.

In a similar way, we can write the upward infrared energy flux at the top of the atmosphere (the top of the upper atmospheric layer), which is the OLR, as

$$OLR = \left(1 - \varepsilon_2\right)\left[\left(1 - \varepsilon_1\right)\sigma_{SB} T_S^4 + \varepsilon_1 \sigma_{SB} T_1^4\right] + \varepsilon_2 \sigma_{SB} T_2^4. \tag{2.5}$$

Finally, we work out the expressions for $LW \downarrow$ at each layer edge, starting at the top of the atmosphere, where $LW \downarrow = 0$. At the interface between the two atmospheric layers, we have

$$\left(LW \downarrow\right)_{3/2} = \varepsilon_2 \sigma_{SB} T_2^4. \tag{2.6}$$

The downwelling infrared radiation at the Earth's surface is

$$\left(LW \downarrow\right)_S = \left(1 - \varepsilon_1\right)\varepsilon_2 \sigma_{SB} T_2^4 + \varepsilon_1 \sigma_{SB} T_1^4. \tag{2.7}$$

Equations (2.4)–(2.7) can be used to find the upward and downward fluxes of infrared radiation as functions of height in the two-level model, given the various temperatures and emissivities. From these fluxes, we can determine the infrared radiative heating or cooling rate throughout the atmosphere and at the surface. We can also determine the bulk emissivity.

The method outlined above can be extended to any number of layers. It can be generalized to include scattering and to take into account the variations of emission, absorption, and scattering with wavelength. It can be adapted to describe solar radiation as well as terrestrial radiation. For further information, see the excellent review article by Pierrehumbert (2011).

Having derived (2.4)–(2.7), we should now use them for something. Here is a very simple and highly idealized example of how they can be applied. We will assume that the two atmospheric layers are in energy balance and use this assumption to calculate the temperatures of the layers in terms of the surface temperature. For simplicity, we assume that *only* longwave radiation influences the atmospheric temperature: this means that the atmosphere is transparent to solar radiation, the air is completely dry (think "desert planet") so that latent heat release does not occur, and there is no energy transport by the winds. These are pretty drastic simplifications, and in combination with our assumption of energy balance they imply that the atmospheric column is in "longwave radiative equilibrium." In such a case, the net (up minus down) infrared flux has to be constant with height, all the way from the surface to the top of the atmosphere. For our two-layer model, we can impose this condition by setting

$$\sigma_{SB}T_S^4 - (LW\downarrow)_S = (LW\uparrow)_{3/2} - (LW\downarrow)_{3/2}, \text{ and}$$
$$(LW\uparrow)_{3/2} - (LW\downarrow)_{3/2} = OLR. \tag{2.8}$$

After substitution from (2.4)–(2.7), these two equations can be used to solve for the temperatures of the two model layers. The results are

$$\sigma_{SB}T_1^4 = \left[\frac{2+(1-\varepsilon_1)\varepsilon_2}{4-\varepsilon_1\varepsilon_2}\right]\sigma_{SB}T_S^4 \text{ and}$$
$$\sigma_{SB}T_2^4 = \left(\frac{2-\varepsilon_1}{4-\varepsilon_1\varepsilon_2}\right)\sigma_{SB}T_S^4. \tag{2.9}$$

The first of these is valid provided that $\varepsilon_1 \neq 0$, and the second provided that $\varepsilon_2 \neq 0$. By trying some numerical values for ε_1 and ε_2, and choosing a surface temperature, you can calculate the temperatures T_1 and T_2. For example, putting T_S equal to the observed globally averaged value of 288 K, and setting $\varepsilon_1 = \varepsilon_2 = 0.5$, we find that $T_1 = 253$ K and $T_2 = 229$ K. These are comparable to the observed temperatures in the middle and upper troposphere, respectively.

Substituting from Equation (2.9) back into Equation (2.5), we find, after some algebra, that

$$OLR = \left\{\frac{(2-\varepsilon_1)(2-\varepsilon_2)+2\varepsilon_1\varepsilon_2}{4-\varepsilon_1\varepsilon_2}\right\}\sigma_{SB}T_S^4. \tag{2.10}$$

Comparing Equation (2.10) to Equation (2.3), we see that the expression in curly braces, in Equation (2.10), is the bulk emissivity. Using $\varepsilon_1 = \varepsilon_2 = 0.5$, we obtain a bulk emissivity of 0.73. Many additional results can be worked out, but we will stop here.

PLANETARY ENERGY BALANCE

As mentioned in Chapter 1, the globally and annually averaged OLR is very nearly equal to the globally and annually averaged solar radiation absorbed by the Earth. Both numbers are close to 240 W m^{-2}, which, summed over the area of the Earth, adds up to about 10^{17} W. The net flow of radiation across the top of the atmosphere, per unit area, is given by

$$N \equiv \frac{1}{4}S(1 - \alpha) - \varepsilon_B \sigma_{SB} T_S^4. \qquad (2.11)$$

If the Earth is in energy balance, $N = 0$. Observations are consistent with $N = 0$, in the sense that zero lies within the error bars on the observed value of N (Loeb et al., 2009). It is believed that, at present, the true value of N is positive and somewhat less than +1 W m^{-2}, which would mean that the OLR is slightly less than the absorbed solar radiation. Unfortunately, measurements are not yet accurate enough to test this directly. We return to this topic in Chapter 8.

CLIMATE CHANGE ON THE BACK OF AN ENVELOPE

With this preparation, we can take a first look at how a change in the flow of radiation can affect the climate. The analysis is highly simplified, but it illustrates some basic concepts.

Suppose that the various quantities that appear in (2.11) are perturbed by some external change, such as a brightening of the Sun, or by an internal fluctuation, such as a change in the planetary albedo. We find that

$$\Delta N \cong \frac{1}{4}\Delta S(1 - \alpha_0) - \frac{1}{4}S_0\Delta\alpha - 4\varepsilon_0\sigma T_0^3\Delta T_S - \Delta\varepsilon\sigma T_0^4. \quad (2.12)$$

In (2.12), quantities with subscript zero represent the unperturbed state, ΔS is the perturbation to S, and ΔT_S, $\Delta\varepsilon$, and $\Delta\alpha$ are defined similarly. The simplest way to get (2.12) from (2.4) is to use the product rule for differentiation; here we apply that rule with finite perturbations, so the result is approximately valid if the perturbations are small enough.

We analyze a change in the *equilibrium* state of the system; this means that the system is assumed to be in energy balance both before and after the perturbation, so that $\Delta N = 0$. We start from an equilibrium, we perturb it, things evolve, and we arrive at a new equilibrium. Then (2.12) can be written as

$$0 \cong \frac{1}{4}\Delta S(1 - \alpha_0) - \frac{1}{4}S_0\Delta\alpha - 4\varepsilon_0\sigma T_0^3\Delta T_S - \Delta\varepsilon\sigma T_0^4. \quad (2.13)$$

Using $\frac{1}{4}S_0(1 - \alpha_0) = \varepsilon_0\sigma T_0^4$, which is a statement of the assumed equilibrium of the starting state, we can rearrange (2.13) to

$$\frac{\Delta T_S}{T_0} \cong \frac{1}{4}\left(\frac{\Delta S}{S_0} - \frac{\Delta\alpha}{1 - \alpha_0} - \frac{\Delta\varepsilon}{\varepsilon_0}\right). \quad (2.14)$$

Equation (2.14) relates the fractional change in surface temperature to the fractional changes in solar output, the planetary albedo, and the bulk emissivity. The factor of 1/4 on the right-hand side of Equation (2.14) can be traced back to the fourth power of the temperature on the right-hand side of Equation (2.11).

As a first example, suppose

$\Delta \alpha = 0$, that is, there is no change in the planetary albedo
$\Delta \varepsilon = 0$, that is, there is no change in the bulk emissivity

Then (2.14) reduces to

$$\frac{\Delta T_S}{T_0} \cong \frac{1}{4} \frac{\Delta S}{S_0} \qquad (2.15)$$

This says that the fractional change in surface temperature is equal to one-fourth of the fractional change in solar output. From measurements, we know that the globally averaged surface temperature is currently 288 K. Using this value for T_0, we find that a 1% change in solar output will lead to a 0.72 K change in surface temperature. Observed fluctuations of solar output over the past 30 years are about 0.1% in magnitude, so the expected temperature variations of T_S due to changes in the Sun are less than one-tenth of a kelvin.

As a second example, consider a simplified "global warming" case in which

$\Delta S = 0$, that is, there are no changes in the Sun's output

$\Delta\alpha = 0$, that is, there is no change in the planetary albedo

Then (2.14) reduces to

$$\frac{\Delta T_S}{T_0} \cong -\frac{1}{4}\frac{\Delta\varepsilon}{\varepsilon_0}. \tag{2.16}$$

Equation (2.16) says that the change in the surface temperature is entirely due to a change in the bulk emissivity. The minus sign appears because a decrease in ε makes it harder for the Earth to emit infrared, and so leads to a warming. An increase in atmospheric CO_2 leads to a decrease in the bulk emissivity. It is known from the measured optical properties of CO_2 that, for the current climate, a doubling of CO_2 relative to its preindustrial concentration would reduce the OLR by 4 W m^{-2}, so that $\Delta\varepsilon\sigma T_0^4 \cong -4$ W m^{-2}. We also know, from satellite observations, that the OLR is $\varepsilon_0\sigma T_0^4 = 240$ W m^{-2}. Forming the ratio, we find that $-\left(\frac{\Delta\varepsilon}{\varepsilon_0}\right) = \frac{4}{240} \cong 0.017$. This means that doubling CO_2 creates a 1.7% perturbation to the OLR.

You should think of this reduction in the OLR as happening "instantaneously"; imagine that we could somehow double CO_2 without changing the temperature, or the albedo, or anything else. The system would then evolve so as to make the OLR increase again, reestablishing global energy balance. It would accomplish this by warming up, that is, by increasing T_0. Using the current globally averaged surface temperature of 288 K, we find from (2.13) that doubling CO_2 leads to a 1.2 K increase in the surface temperature—less than a 0.5% warming.

To repeat, the simple analysis above suggests that a 1.7% perturbation of the top-of-the-atmosphere radiation will lead to a 0.5% change in the surface temperature. Nothing outrageous there. The result was obtained using very simple ideas, which could be compactly summarized on the back of an envelope.[7]

In the preceding analysis, we explicitly assumed no changes in the planetary albedo. A second assumption, unstated until now, was that the bulk emissivity changes *only* due to increasing CO_2. For reasons discussed in Chapter 6, both of these assumptions are highly questionable. Relaxing them leads to a prediction of stronger warming in response to a doubling of CO_2.

Many people feel intuitively that the climate system will "adjust" to an increase in atmospheric CO_2, so as to maintain a balance. The analysis above shows that they are right: the system adjusts by warming.

POLEWARD ENERGY TRANSPORTS BY THE ATMOSPHERE-OCEAN SYSTEM

Next, we examine the variations with latitude of the solar and infrared radiative fluxes at the top of the atmosphere and their implications for poleward energy transport by the atmosphere and oceans. Another bit of terminology: changes with latitude are called "meridional" changes because they occur along meridians, which are lines of constant longitude. Figure 2.5 shows measurements of the meridional variations of the annually and zonally averaged absorbed solar radiation, OLR, and net

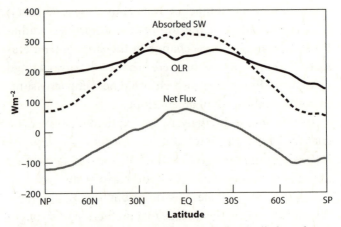

Figure 2.5. The zonally and annually averaged absorbed solar radiation (dashed), outgoing longwave radiation (solid black), and net radiation at the top of the atmosphere (gray), as observed from satellites. The data are discussed by Wielicki et al. (1996).

radiation (absorbed solar minus OLR) at the top of the atmosphere, as observed in the Earth Radiation Budget Experiment (ERBE; Barkstrom et al., 1989).

The absorbed solar radiation is roughly symmetrical about the Equator. It has a slight minimum near the Equator, caused by many high bright clouds. The zonally averaged OLR varies less strongly with latitude. Generally, the warm tropical regions emit more and the cold polar regions emit less, but the OLR actually has its maximums in the subtropics, on either side of the Equator, where hot deserts emit upward under dry, clear skies. The OLR is relatively small over the cold poles, but it also has a local minimum in the warm tropics due to the trapping of terrestrial radiation by the cold, high tropical clouds and by water vapor. The net radiation into the atmosphere is positive in the tropics and negative in higher latitudes. This means that the tropics absorb more radiation than they emit and the polar regions emit more than they absorb. Energy is conserved, so there must be a flow of energy inside the system, from the tropics toward the poles.

This energy flow can be calculated from the measurements of the zonally and annually averaged net top-of-the-atmosphere radiation, shown by the solid curve in Figure 2.5. The details of the calculation are straightforward but will not be explained here. The results are shown in Figure 2.6. The curve is very smooth, and it has a simple shape. Observed quantities rarely vary in such a simple way! A poleward energy flow occurs in both hemispheres, with positive values in the Northern Hemisphere, negative values in the Southern Hemisphere, and a zero very close to the Equator.

The energy flow collapses to zero, *exactly*, at both poles. This has to be true. To see why, think of the poles as literal poles, like flag poles, but with infinitesimal diameters. Such poles would have zero mass (they are "points"), and therefore zero capacity to hold energy. The energy flow has to go to zero at the poles simply because it is impossible for a finite amount of energy per unit time to flow into or out of such "points" of zero mass.

The winds and ocean currents accomplish the poleward energy transports shown in Figure 2.6. If the transport of energy from place to place by the atmosphere and

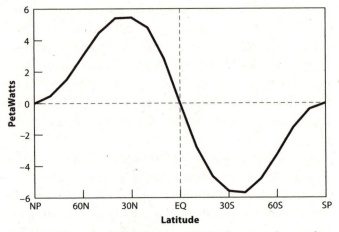

Figure 2.6. The poleward energy transport by the atmosphere and ocean combined, as inferred from the observed annually averaged net radiation at the top of the atmosphere.

A petawatt is 10^{15} W.

oceans could somehow be prevented, then each part of the Earth would have to come into *local* energy balance by adjusting its temperature, water vapor, and cloudiness so that the OLR *locally* balanced the *locally* absorbed solar radiation. Such a locally balanced climate state would be very different from the one we actually see. The tropics would be much hotter and the poles would be much colder. The circulations of the atmosphere and oceans have a moderating effect on the global distribution of temperature, tending to warm the higher latitudes and cool the tropics. As we will see in Chapter 4, however, these same thermal contrasts drive the global circulation of the atmosphere.

The discussion above may give the impression that the poleward energy transport by the atmosphere and oceans is *caused* by the Equator-to-pole differences in the net radiation at the top of the atmosphere. The reality is more complicated because the distributions of the absorbed solar radiation and the OLR are influenced by weather systems. For example, weather systems are associated with clouds, which affect both the absorbed solar radiation and the OLR. For this reason, it is not right to think of the meridionally varying net top-of-the-atmosphere radiation as an externally imposed forcing function; the net radiation is strongly influenced by the circulation itself.

RADIATIVE EXCHANGES BETWEEN THE SURFACE AND THE ATMOSPHERE

Although Earth as a whole is very close to radiative energy balance, the atmosphere and surface are not. The net infrared radiation at the surface is upward, so it acts to cool the surface and warm the atmosphere. The net infrared cooling of the surface is more than compensated by the absorbed solar radiation, however, so that the surface experiences a net radiative energy gain, listed in Table 2.1 as 98 W m^{-2}. This net radiative warming of the surface is balanced by upward turbulent heat fluxes from the surface into the atmosphere, as discussed in Chapter 3.

The globally averaged radiative energy budget of the atmosphere is shown in Table 2.2. The combination of

Table 2.2
The globally and annually averaged surface radiative energy budget of the atmosphere. A positive sign means that the atmosphere is warmed.

Absorbed solar radiation (239 – 161)	78 W m^{-2}
Net emitted terrestrial radiation (−239 + 63)	−176 W m^{-2}
Net radiative heating	−98 W m^{-2}

(net up minus down) infrared at the surface and the upward infrared at the top of the atmosphere acts to cool the atmosphere at the rate of 176 W m^{-2}. This infrared cooling of the atmosphere is strong enough to win out over the warming by absorbed solar radiation, which is only about 78 W m^{-2}. The net effect of all radiation flows is, therefore, to cool the atmosphere at the rate of 98 W m^{-2}. Recall that that is also the rate at which the surface is radiatively warmed.

An interpretation is that there is a net flow of radiative energy from the atmosphere to the surface. Both solar and infrared radiation contribute strongly to this flow. It must be balanced by an *nonradiative* energy flow from the surface back into the atmosphere. Chapter 3 explains how that works.

3 HOW TURBULENCE AND CUMULUS CLOUDS CARRY ENERGY UPWARD

....

ENERGY FLOWS BACK TO THE ATMOSPHERE

AS DISCUSSED IN CHAPTER 2, THE LAND SURFACE AND ocean receive a net input of solar and infrared radiation at the average rate of 103 W m^{-2}. The temperature of the land surface is easily warmed by this energy input because the energy cannot penetrate quickly down into the soil column. On daily time scales, the Sun warms just the top few centimeters of the soil. Energy flows back into the atmosphere from the rapidly warming land surface, through mechanisms that are discussed below.[1]

The ocean is a different story. A column of ocean water can accumulate and hold a lot of the incident radiative energy on daily time scales, for two reasons: First, sunlight can penetrate many meters into the water column. Second, turbulence in the upper ocean quickly mixes the energy vertically through a depth that varies from a few meters to a few hundred meters.

In addition, ocean currents can carry the absorbed radiative energy over large horizontal distances. This energy transport by the currents means that, locally, there can be a net flow of energy into or out of the sea surface, even on annual or longer time scales.

....

Most of the solar and infrared energy input to the oceans and land surface is used to evaporate water. The resulting upward latent heat flux cools the surface, at the globally averaged rate of 80 W m^{-2}, and adds the same amount of energy to the atmosphere, in the form of latent heat. Atmospheric warming occurs later, when the vapor condenses, often far away in space and time from where the surface evaporation occurred. Not surprisingly, most evaporation occurs over the oceans, but even over land the water vapor flux is almost always upward. A familiar exception is the weak and short-lived downward flux of vapor that occurs when dew or frost forms on a chilled terrestrial surface.

A smaller radiatively driven energy flow from the Earth's surface to the atmosphere is the "surface sensible heat flux." The adjective "sensible" is apt because the energy can be felt; it is the thermal energy associated with the temperature of the air. The atmosphere acquires sensible heat when cool air comes into contact with the warmer surface. The air and the surface exchange energy through molecular conduction and tend to approach the same temperature. The surface cools off, the air warms up, and the result is a net transfer of sensible energy from the surface to the atmosphere. Globally, the surface sensible heat flux is about 17 W m^{-2}, less than a quarter as large as the surface latent heat flux.[2]

Although the globally and annually averaged latent and sensible heat fluxes are upward, the instantaneous local fluxes vary a lot and can be either upward or downward. The sensible heat flux is weakly upward over most of the oceans, worldwide, and it is usually strongly upward over sun-warmed land during the day. At night, the sensible heat flux is usually downward over land because infrared emission cools the ground more quickly than the air. The nocturnal downward flux of sensible heat is almost always small because warm air has to sink (buoyancy works against that) and cold air has to rise (same issue). Conversely, an upward sensible heat flux is promoted by buoyancy. Strong upward sensible heat fluxes occur not only over heated land but also near coastlines in winter, when very cold air flows quickly out over warm water.

The upward fluxes of latent and sensible energy penetrate, within minutes, through the depth of the atmospheric boundary layer. How do those energy fluxes actually work? One way to transport energy through the atmosphere is with a smooth, uniform flow of air (horizontal or vertical) that actually carries mass (air) from place to place. A smooth, uniform upward flow of air is capable of changing the distributions of temperature and moisture with height.[3] That process is called upward "advection," or sometimes "large-scale advection." The word comes from a Latin source meaning "carry to."

There is a second way that the winds can change the sounding. The atmosphere contains a variety of small-scale motions that significantly affect the global-scale circulation. These include turbulence in the boundary layer near the Earth's surface, the updrafts and downdrafts associated with cumulus clouds, turbulence above the boundary layer in clouds and in regions of strong shear, and atmospheric waves excited by a variety of mechanisms including strong winds blowing over mountains. These can be lumped together under the generic heading "small-scale eddies."

Think of a sunny summer afternoon over land, when the air next to the ground is strongly heated. This is analogous to adding heat to a pan of water on a stove. In such a pan, a turbulent churning motion develops, as warm water rises away from the hot bottom and is replaced by sinking, cooler water that has been in contact with the cool air above. In a similar way, the upward sensible and latent energy fluxes from the Earth's surface create buoyant, humid patches of air. Warm thermals, with high energy content, break away from the surface, somewhat randomly, and float upward under the influence of buoyancy. In compensation, air with less energy moves downward, around and between the thermals, to fill the gaps that they leave behind. Many thermals can be generated, leading to a fluid dynamical commotion.

The net effect of the warm thermals floating upward and the cooler air sinking around them is that, over time, the upper levels are warmed and the lower levels are cooled. There is a net upward "eddy" flux of energy. Here the word "eddy" refers to the round-trip circulations associated with the rising thermals and the sinking air around them. Eddy fluxes are due to small-scale advection, but they produce no "net" or large-scale mass flow. That is how they differ from large-scale advection. Eddy fluxes are an important way that small scales (the eddies) can influence larger scales. They are a key mechanism for scale interactions. Further explanation is given in the appendix to this chapter.

Eddy fluxes have important effects on the global circulation of the atmosphere. They can be due to thermals, as discussed above, but they also arise in other ways. The most important examples are the following:

Vertical fluxes due to turbulence, especially in the boundary layer, including the turbulence associated with thermals as in the discussion above

Vertical fluxes due to cumulus clouds,[4] discussed later in this chapter

The vertical flux of momentum due to small-scale waves, which exert their effects mainly in the stratosphere and above, and which will not be discussed in this book

"Convection," which comes from a Latin source meaning "carry with," is the term used by atmospheric scientists to denote a general buoyancy-driven overturning of the air, with warm air rising and cooler air sinking.[5] Decades ago, the longer expression "natural convection" was used to mean the same thing. Convection can occur in clear air, as often happens in the boundary layer, and it can also occur in clouds. In fact, convection is perhaps the single most important process leading to the formation of clouds.

Under conditions that will be explained later in this chapter, some of the thermals that grow near the surface can break away from the boundary layer and grow into

Figure 3.1. Top panel: A group of small and large cumulus clouds.

Rapidly rising moist air forms the cloudy towers. The air around the towers is slowly sinking. Some of the larger towers have flat cloud layers near their tops. These are regions where the air has stopped rising and is flowing out to the sides.

Source: This beautiful photo was taken by Marco Lillini, an airline pilot, and is used with his kind permission. His web site is here: http://www.lillini.com/.

Bottom panel: A full-disk image of the Earth, in visible light, taken on November 24, 2010.

North America can be seen at the upper right. The Equator is directly below the satellite that took the image. The Intertropical Convergence Zone (ITCZ) appears as a "white stripe" that is oriented east-west, slightly north of the Equator. At any given time, the ITCZ contains many thousands of small and large cumulus clouds, comparable to those shown in the upper panel of the figure.

Source: NASA-Goddard Space Flight Center, data from NOAA GOES. See "Pacific ITCZ" at http://goes.gsfc.nasa.gov/text/goes11results.html.

thunderstorms, like the ones shown in the upper portion of Figure 3.1. A typical thunderstorm is 5 to 10 km across. In the atmosphere as a whole, many thousands of thunderstorms are occurring at any given moment. The storms grow rapidly upward because they contain strong, organized updrafts, with speeds of 20 m s^{-1} or more, much faster than any elevator you have ever ridden on. The updrafts lift energy through the depth of the troposphere, and sometimes even stab into the lower stratosphere. They also lift moisture, momentum, and various chemical species. The condensation of water vapor in thunderstorms leads to the formation of large water drops and snowflakes and produces much of the precipitation that falls on the Earth's surface. A thunderstorm's updrafts eject smaller, free-floating liquid and ice particles, which spread out to form flat, "stratiform" clouds that cover large areas and therefore have major effects on the flows of both solar and infrared radiation.

Thunderstorms are particularly common in the tropics. The Intertropical Convergence Zone (ITCZ; shown in the lower panel of Figure 3.1) is a stormy belt that roughly follows a circle of constant latitude, slightly north of the Equator. It is an important element of the climate system and will come up again in later chapters of this book. At any given moment, the ITCZ contains many thousands of cumulus clouds like the ones shown in the upper panel of Figure 3.1. The cumuli collectively lift energy toward the tropopause at the enormous rate of about 3×10^{16} W, roughly a third as large as the solar radiation absorbed (or the infrared radiation emitted) by the whole Earth. Thunderstorms also occur over warm continents, in tropical cyclones (colloquially known as hurricanes and typhoons) over the oceans, in winter storms, and in various other weather systems.

Despite their relatively small size, cumulus clouds are major players in the energy budget of the global climate system.

But the story starts in the boundary layer.

TURBULENT MIXING

In Chapter 1, the boundary layer was described as turbulent and strongly affected by direct contact with the Earth's surface. The boundary layer is actually created by turbulence and is the direct recipient of the turbulent fluxes of latent and sensible energy that flow upward into the atmosphere. The energy gradually accumulates in the boundary layer, like water building up behind a dam, waiting for release.

The upward sensible heat flux helps to maintain the turbulence in the boundary layer, by creating the thermals described earlier. Turbulence is familiar in everyday life, but it is not easy to define. You know it when your plane flies through it. A turbulent flow can be described as random, disordered, noisy, and complicated.

Over warm surfaces, the turbulent eddies in the boundary layer often take the form of thermals, plumes, and vortices. The largest ones are typically a few kilometers wide. Dust devils are the visible manifestations of particularly strong rotating plumes over sun-baked, arid landscapes. Physically similar but invisible and mostly weaker plumes are routinely generated over heated terrain and also over the warm oceans, especially when the air is much colder than the surface water.

Weaker turbulence of a different type creates a boundary layer during the night over the cool land surface. Nocturnal turbulence is created by wave-like instabilities of the wind.

One of the effects of strong turbulence is that it mixes things up. A familiar example is the mixing produced by turbulence in a blender. Mixing can make some things homogeneous, like the uniform taste and color of your favorite blended beverage. Thinking of the turbulent boundary layer as a giant blender, we can ask, what properties of the air are homogenized by the blender? What does the blender blend?

Some quantities can be homogenized by sufficiently strong mixing, and some can't. To see why, consider a "parcel" of air, that is, a quantity of air (mass) small enough that it can be considered to move with a well-defined velocity, and to be characterized by a single pressure, temperature, and humidity, but large enough that it contains a huge number of molecules. One gram of air will do. The time rate of change of the mass fraction of oxygen in the parcel satisfies

$$\frac{DO_2}{Dt} = 0. \tag{3.1}$$

Here the symbol $\frac{D}{Dt}$, which is called the "Lagrangian time derivative," denotes the time rate of change *following a (possibly) moving parcel of air*. The Lagrangian time derivative is useful because it allows us to describe what happens to a parcel that is moving around in the atmosphere. In the case of Equation (3.1), the answer is simple:

the mass fraction of oxygen does not change as the parcel moves around. We say that the mass fraction of oxygen is "conserved," and that O_2 is a "conservative" variable.

Mixing homogenizes conservative variables. The oxygen concentration is homogenized by mixing, and so are the mass fractions of nitrogen, argon, neon, and carbon dioxide, because in the Earth's atmosphere all of those gases are chemically inert and do not undergo phase changes. That is *why* their mass fractions are conserved following parcels.

In contrast, the blender does not and cannot blend temperature, no matter how strong the mixing becomes. To see why, take a look at the "thermodynamic energy equation," which describes how the temperature of a parcel changes. It can be written like this:

$$\rho \frac{D}{Dt}\left(c_p T\right) = \frac{Dp}{Dt} + LC + Q_{rad}. \tag{3.2}$$

Here c_p is the constant specific heat of air at constant pressure; its value is 1,004 J K^{-1} kg^{-1}. The product $c_p T$ is called the enthalpy. The latent heat of condensation, per unit mass, is represented by L. The actual condensation rate is represented by C. Its units are grams of water condensed per gram of dry air per unit time, so its dimensions are simply "per unit time." The product LC represents the rate at which latent heat is released, per unit mass, when water vapor condenses. The radiative heating rate per unit mass is Q_{rad}. These two terms together comprise the "heating rate," and of course negative heating is called cooling. Throughout most of the atmosphere, most of the time, Q_{rad} and LC are zero or small compared to the other two terms of (3.2), which therefore nearly balance.

Equation (3.2) tells us that the temperature is *not* conserved. It can change due to heating or cooling, which is not surprising, and also due to changes in the pressure acting on the parcel, which occur when $\frac{Dp}{Dt} \neq 0$. If the parcel moves upward, it goes to a place where the pressure is lower, and so $\frac{Dp}{Dt} < 0$. Equation (3.2) tells us that the pressure fall will produce a temperature decrease. We say that the parcel is cooled by expansion because the volume of the parcel literally gets bigger as the pressure acting on it decreases. Conversely, if the parcel moves downward, the increased pressure compresses it (reducing its volume), and so its temperature increases. We say that the parcel has been warmed by compression. No matter how vigorously the turbulence mixes the air, the temperature will tend to decrease upward, in the direction of lower pressure, and increase downward, in the direction of higher pressure, because of expansion and compression. *That's why the temperature can't be homogenized by mixing*, even when no energy is added or removed.

Conservative variables are easy to work with because they obey simple rules; again, compare the simple Equation (3.1) to the more complicated Equation (3.2). It would be nice if we could define a conservative (or nearly conservative) variable that is closely related to the temperature. It turns out that there are actually several ways to do so. The temperature-like conservative variable that we will use in this book is the "dry static energy," which is denoted by s and defined by

$$s \equiv c_p T + gz. \tag{3.3}$$

Here the enthalpy, $c_p T$, is added to the gravitational potential energy per unit mass, which is given by gz, where g is the downward acceleration due to gravity.

When a parcel moves upward, it cools due to expansion, as described above, but at the same time its gravitational potential energy increases. It turns out that the temperature and potential energy changes nearly cancel, so that the sum of the two, which is the dry static energy, (almost) doesn't change, as long as there is no energy source or sink due to heating or cooling. This fact is expressed by

$$\rho \frac{Ds}{Dt} = LC + Q_{rad}. \tag{3.4}$$

The terms on the right-hand side of (3.4) represent the heating due to condensation and radiation, respectively. Compare (3.4) to (3.2). Equation (3.4) is simpler.

In the absence of heating, dry static energy is a conservative variable, and so it can be homogenized by mixing. As discussed later, when the air is cloudy, so that $LC \neq 0$, the effects of latent heat release prevent the dry static energy from being homogenized by mixing.

Turbulence can vertically homogenize the dry static energy through the depth of the boundary layer, unless a cloud forms. Strong turbulence also homogenizes water vapor, unless a cloud forms. More specifically, the turbulence homogenizes the *mass fraction* or concentration of water vapor. The atmospheric science jargon for the mass fraction of water vapor is the "specific humidity," which is defined by $q \equiv \rho_{vapor}/\rho$, where ρ and ρ_{vapor} are the total density and the density of water vapor, respectively. The specific humidity is typically a few parts per thousand. It satisfies

$$\rho \frac{Dq}{Dt} = -C, \tag{3.5}$$

where C is the rate at which vapor is condensing into liquid, per unit mass of air. A negative value of C signifies the evaporation of liquid to form vapor. For example, when falling rain evaporates into subsaturated air, $C < 0$. Where $C = 0$, which basically means away from clouds, water vapor is conserved, and it can be homogenized by mixing.

Over huge expanses of the tropical oceans, turbulent mixing ensures that both the dry static energy and the mixing ratio of water vapor are (almost) vertically uniform through the boundary layer, which in those regions typically occupies the atmosphere's lowest 500 or 600 m.

STRATIFICATION

Above the boundary layer, turbulence is nearly absent except inside clouds and in isolated patches of "clear-air turbulence," which are often associated with breaking waves. Because the turbulence is weak, mixing is overwhelmed by heating and other processes, and even

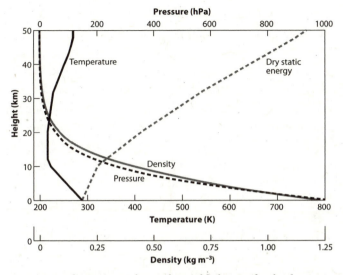

Figure 3.2. Like Figure 1.1, but with an added curve for the dry static energy, divided by c_p to give units of temperature, and with a modified temperature scale to accommodate the large values of s/c_p in the stratosphere.

The dry static energy increases upward through most of the troposphere, and it increases even more strongly upward in the stratosphere. The figure is based on the U.S. Standard Atmosphere.

conservative variables can change rapidly with height. An example is the upward increase of dry static energy shown in Figure 3.2, which is based on the U.S. Standard Atmosphere. The figure shows that the dry static energy varies strongly with height.[6]

The air 10 km above the surface is pretty cold: about 220 K, which is −53°C. If a parcel of that air descends to the surface, its gravitational potential energy decreases by $g * 10{,}000$ m, or about 100,000 J kg^{-1}. Conservation of dry static energy then implies that it will warm by about 100 K, to a temperature of near 320 K, which is hot by human standards, about 30 K warmer than the air at the surface! What this example illustrates is that the air near the tropopause is actually "warmer," in terms of dry static energy, than the air near the surface. In Chapter 1, I asked, "If buoyancy pushes warm air up and cold air down, then shouldn't we find the warm air on top and the cold air below?" Now you see that *the warm air actually is on top*, if "warm" is interpreted to mean high dry static energy.

There are several reasons why the dry static energy increases upward. For the troposphere, one explanation is that we are seeing the end result of a process in which the warm air rises to the top. A second, reinforcing reason is that, as discussed later in this chapter, the rising air often experiences an increase in its dry static energy due to latent heat release as a cloud forms. Similarly, sinking air can be cooled, especially by radiation. Equation (3.4) tells us that radiative cooling ($Q_{rad} < 0$) will cause the dry static energy of a parcel to decrease with time. Radiative cooling of sinking parcels leads to a downward decrease—or upward increase—of the dry static energy. Finally, for the stratosphere, the strong upward increase of the dry static energy is due to the absorption of solar UV by ozone, which was discussed earlier.

The atmosphere is like a cake in which the layers are arranged in the order of their dry static energies, with the largest dry static energy on top and the smallest at the bottom. We say that the atmospheric column is stratified.

STATIC STABILITY AND INSTABILITY IN DRY AIR

Because of buoyancy forces, *there is a tendency for parcels to maintain their relative vertical positions in a stratified column of air.* In other words, vertical motions are inhibited by stratification.

We say that a stratified column of air is statically stable. The concepts of stability and instability are probably familiar to you. They are very important in atmospheric science and will come up again later in this book. The classical concept of stability is based on the analysis of a system *in equilibrium*, and the response of such a system to small perturbations. A stable system is one in which small perturbations are opposed by a "restoring force." In the case of static stability, the restoring force is buoyancy.

When instability occurs, perturbations are amplified. An elementary example of an unstable system is a pencil precariously balanced, in equilibrium, on its point. A small perturbation of the pencil's position is amplified by gravity, causing the pencil to fall over.

For the atmosphere, one relevant equilibrium is an idealized, horizontally uniform column of air that is in the hydrostatic balance discussed in Chapter 1. Such an idealized equilibrium is often called a "basic state." Because the basic state is horizontally uniform, the temperature of every parcel is the same as the temperature of the neighboring parcels at the same height. Therefore, the buoyancy force acting on a parcel in the basic state is zero. The relevant perturbations, which may or may not be stable, are vertical displacements of the parcels.

A basic state in which the dry static energy *increases upward* is said to be *stably stratified* because vertical motions in such a column are resisted by buoyancy. If a parcel in equilibrium is displaced either upward or downward, buoyancy pushes it back toward its starting point. This is illustrated in Figure 3.3. The buoyancy force arises because the parcel conserves its dry static energy as it moves up or down, in the presence of stratification. If the parcel moves upward, it finds itself surrounded by air with a larger dry static energy.

We refer to the air surrounding the parcel at the same level as the *environment* of the parcel. Because the upward-displaced parcel has a smaller dry static energy than its environment, its temperature is also cooler, and so a downward buoyancy force pushes it back toward its level of origin. It will arrive there with a downward velocity, so it will actually overshoot its level of origin and move further downward, where its environment has a smaller dry static energy. Buoyancy then pushes the parcel back up, it overshoots again, and an oscillation ensues. This is the basic mechanism responsible for a simple type of atmospheric wave motion, which is called a "gravity wave" because of the key role of the buoyancy force.[7] Gravity waves commonly occur in the atmosphere and arise from many causes, including flow over mountains.

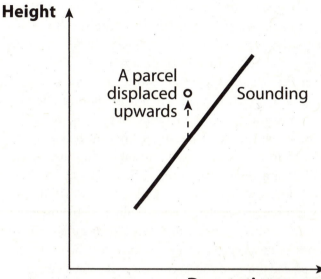

Figure 3.3. Sketch illustrating a parcel that is displaced upward in a statically stable sounding, for which the dry static energy increases upward.

The parcel is assumed to conserve its dry static energy.

In summary, the preceding analysis shows that when the dry static energy of the basic state increases upward, vertical displacements of the air are opposed by buoyancy; the hydrostatically balanced basic state is stable and capable of supporting gravity waves.

This analysis is based on two assumptions: (a) the dry static energy increases upward in the basic state and (b) the parcel conserves its dry static energy as it moves. Both assumptions can break down.

The first assumption fails, for example, over a dry surface that is being strongly heated by the Sun. The resulting strong upward surface sensible heat flux can cause the area-averaged dry static energy to *decrease* upward near the surface, say in the lowest 100 m or so. In fact, that is what usually happens on a summer afternoon over a dry surface like a desert or a parking lot. A basic state in which the dry static energy decreases upward is said to be gravitationally unstable. The instability is called "gravitational" because it is caused by the buoyancy force, which accelerates the upward-drifting thermals discussed earlier. It is also called "static instability" because, unlike some other types of instability, it can work even when the air is not moving. Parcels that are displaced upward, conserving their dry static energy, are warmer than the air around them, so a positive buoyancy force amplifies their initial displacement. Similarly, parcels that are displaced downward are accelerated downward by a negative buoyancy force. The kinetic energy of the thermals comes from the gravitational potential energy of the basic state.

Static instability gives rise to a thermal-filled turbulence, which tends to mix the dry static energy vertically. This is important because if the dry static energy becomes vertically uniform, the instability is eliminated. In other words, the thermals that develop under static instability tend to modify the sounding in such a way that the modified sounding is no longer unstable. Despite this, the instability can be maintained over time, if the surface heating is strong enough. This idea, that the response to instability tends to remove the instability, will come up again later.

The second assumption used in the analysis above, that is, dry static energy conservation, breaks down when water vapor condenses. If conditions are right, condensation can enable an instability that leads to the formation of cumulus clouds.

CUMULUS INSTABILITY

Cumulus instability is a simple, fascinating, and very important process that occurs quite commonly in many parts of the world. Humid air breaks away from the boundary layer, and floats upward under the influence of the positive buoyancy generated through the release of latent heat, just as a hot air balloon can be lofted by a pulse of heat from its burner. The rising, humid air forms a cloudy tower, which can sometimes be many kilometers tall. A tall cumulus cloud precipitates, and sometimes generates lightning and thunder. It is an awe-inspiring thing of beauty.

Before explaining in more detail how cumulus instability works, we need to take a closer look at what happens when water changes phase. Liquid water evaporates when some of the H_2O molecules that make up the liquid move fast enough to escape from the liquid. The fugitive molecules fill the space near the liquid. Some of them will randomly collide with the liquid and merge back into it. Equilibrium occurs when the rate of escape matches the rate of merging back.

In such an equilibrium, the pressure of the vapor increases with the temperature of the liquid because a warmer liquid contains more fast molecules that have the potential to escape. The pressure of the vapor in equilibrium with a neighboring liquid surface is called the "saturation" vapor pressure. The upper panel of Figure 3.4 shows that the saturation vapor pressure of water, denoted by $e^*(T)$, is an exponentially increasing function of the temperature. At the globally averaged surface temperature of the Earth, which is 288 K, $e^*(T)$ increases at the spectacular rate of 7% per kelvin! The shape of $e^*(T)$ can be measured in the laboratory and is well understood theoretically. It has very important implications for the Earth's climate, as will be discussed in Chapters 5 and 6.

The ratio of the actual vapor pressure to the saturation vapor pressure is called the relative humidity. If the relative humidity exceeds 1 (or, as we say, 100%), the vapor begins to condense to form liquid water drops.[8] In practice, this condensation prevents the relative humidity from exceeding 1 by more than a tiny fraction.[9]

Within a millimeter or so of the sea surface, the actual water vapor pressure is close to $e^*(T)$. A few meters above the surface, it is typically about 20% smaller due to turbulent mixing with drier air from aloft. The largest values of $e^*(T)$ occur in the tropics, where the ocean is warm, and are close to 40 hPa. Under those conditions, about 2.5% of the mass of the near-surface air is water vapor.

Earlier, we defined the specific humidity as the ratio of the density of water vapor to the total density, $q \equiv \frac{\rho_{vapor}}{\rho}$. We now define the *saturation* specific humidity, q^*, as

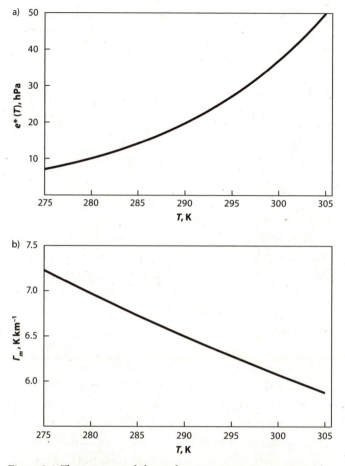

Figure 3.4. The upper panel shows the saturation vapor pressure, e^*, as a function of temperature. The units are hPa, which stands for "hectopascals."

One hPa is 100 pascals. The lower panel shows the moist adiabatic lapse rate, Γ_m, in K km^{-1}, as a function of temperature, and for a pressure of 100,000 Pa, which is close to that at sea level.

the value of the specific humidity for which the partial pressure of water vapor is equal to its saturation value, so that $e = e^*(T)$. The ideal gas law tells us that

$$q^* = \frac{\varepsilon e^*(T)}{p}, \tag{3.6}$$

where $\varepsilon \cong 0.622$ is the ratio of the molecular weight of water vapor to the effective molecular weight of the mixture. Symbolically, we can write $q^*(T, p)$, that is, the saturation specific humidity depends on the temperature and pressure. It does *not* depend on the amount of water that is actually present.

Here is the story of how a cumulus cloud grows: Suppose that a humid but initially unsaturated (and therefore cloud-free) parcel of boundary-layer air is carried upward through the boundary layer by a turbulent eddy. The parcel's dry static energy and specific humidity do not change as it rises, but its temperature decreases. Here comes some more atmospheric science jargon: The rate of decrease of temperature with height is called the "lapse rate." It is denoted by $\Gamma \equiv -\partial T / \partial z$; the minus sign is just a convention that makes the lapse rate positive when the temperature decreases upward, as it usually does in the troposphere. From Equation (3.3), you should be able to see that when the dry static energy does not change with height, the lapse rate is equal to $\Gamma_d \equiv g/c_p$. This particular value is called the "dry adiabatic lapse rate." It is about 10 K per kilometer. A rising parcel that conserves its dry static energy follows the dry adiabatic lapse rate. Because the dry static energy can be vertically homogenized by mixing, the lapse rate in the turbulent boundary layer is often close to the dry adiabatic lapse rate, especially during the daylight hours over land, and over the tropical and subtropical oceans.

As the rising parcel cools, its saturation specific humidity decreases, but its actual specific humidity is conserved. Eventually, if the parcel rises high enough, its saturation specific humidity will decrease to equal its actual specific humidity. At that point, the parcel has reached saturation. We say that the parcel has reached its "lifting condensation level." If the parcel continues to ascend, water vapor begins to condense and latent heat is released. The conversion of vapor to liquid causes the parcel's specific humidity to decrease upward. The release of latent heat causes the parcel's dry static energy to increase upward. Therefore, specific humidity and dry static energy are no longer conserved.

To follow the cumulus formation process further, we need a temperature-like variable that is conserved even when condensation and evaporation are occurring. For this purpose, we can use the "moist static energy," given by

$$h \equiv s + Lq. \tag{3.7}$$

It is the sum of the dry static energy and the latent energy. Recall, however, that the dry static energy is itself the sum of the enthalpy and the gravitational potential energy. Adding Equation (3.4) and L times (3.5), we find that the condensation terms cancel, so that h is governed by

$$\rho \frac{Dh}{Dt} = Q_{\text{rad}}. \tag{3.8}$$

According to (3.8), the moist static energy can be changed by radiative heating or cooling, *but phase changes don't affect it.*[10] The moist static energy is, therefore, "more conservative" than the dry static energy. When the air is not saturated, the dry static energy and water vapor mixing ratio are separately conserved, so their sum, the moist static energy, is again conserved. It is conserved whether or not phase changes are happening.

By putting in some numbers, we can get a feel for the relative sizes of the different energy types. For example, with a temperature of 250 K, which is typical of the midtroposphere, the enthalpy of an air parcel, per unit mass, is about 250,000 J kg^{-1}. The potential energy per unit mass of an air parcel near the tropical tropopause is about 150,000 J kg^{-1}. The latent energy per unit mass of a humid tropical air parcel, near the surface, is about 40,000 J kg^{-1}.

In contrast, for a moderate wind speed of 15 m s^{-1}, the kinetic energy of an air parcel, per unit mass, is a paltry 112 J kg^{-1}. Considering the rate at which energy is absorbed from the sun (about 240 joules per square meter per second, in the global mean), very little atmospheric kinetic energy is actually generated.

We also define the saturation moist static energy, $h^* \equiv s + Lq^*$. The saturation moist static energy is *not* conserved. As with q^*, the value of h^* does not depend on the amount of moisture that is present. The condition that the relative humidity cannot exceed 100% implies that, for any parcel of air, $q \leq q^*$ and $h \leq h^*$. When the air is saturated, the specific humidity and the moist static energy are equal to their respective saturation values.

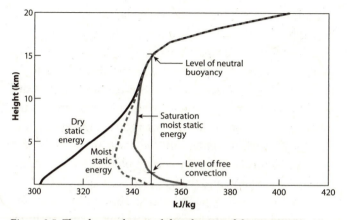

Figure 3.5. The observed vertical distribution of the moist static energy (dashed curve), for January, in kJ kg⁻¹.

A minimum occurs in the tropics, about 5 km above the surface. The vertical profiles of the dry static energy (solid black curve) and saturation moist static energy (solid gray curve) are also shown. The thin vertical line represents the moist static energy of a parcel rising moist adiabatically from near the surface, conserving its moist static energy. The parcel is positively buoyant whenever the thin vertical line is to the right of the dashed line, that is, from about 1 km to 15 km above the surface. Further explanation is given in the text. The plots are based on time averages of data collected during a field experiment called TOGA COARE.

Figure 3.5 shows typical soundings of the dry static energy, moist static energy, and saturation moist static energy for a rainy location over the warm tropical ocean. In the upper troposphere and stratosphere, the moist static energy increases upward. In the tropics the moist static energy decreases upward in the lower troposphere and has a minimum in the middle troposphere. The explanation is simple: above the middle troposphere (or at any altitude near the poles), the water vapor mixing ratio is negligible, so that the moist static energy reduces, approximately, to the dry static energy. Recall that s normally increases upward. Near the surface, where water vapor is plentiful, especially in the tropics, the upward decrease of the water vapor mixing ratio overwhelms the upward increase of s, so that the moist static energy decreases upward. *It follows that the moist static energy has a minimum in the tropical middle troposphere.* This minimum is important, and we will discuss it again later.

Back to our rising parcel. Recall that it has now risen and cooled enough to become saturated. As it continues to rise, more and more of its water vapor condenses. Its moist static energy remains constant with height, however, because the moist static energy is not affected by condensation.

The thin vertical line in Figure 3.5 shows the moist static energy of the rising parcel, starting from the mean moist static energy near the surface. The figure shows that, as the parcel rises, its moist static energy exceeds that of the surrounding, dryer air. The reason is that the cloudy rising air is warmer than its environment, and it also contains more water vapor. Because the rising air has a high moist static energy and the surrounding, sinking clear air has a smaller value, *the net effect is that moist static energy is carried upward.* Cumulus clouds therefore produce an upward energy transport. Later we will come back to this important point.

Although the now-saturated parcel's temperature continues to decrease with height, it does so more slowly than before because of the release of latent heat. Rather than following the strong dry adiabatic lapse rate of 10 K km^{-1}, *the parcel follows the more gradual "moist adiabatic lapse rate," denoted by* Γ_m. The moist adiabatic lapse rate can be calculated directly from the thermodynamic properties of water vapor; the details are omitted here for simplicity. The value of Γ_m varies with temperature and pressure. It is plotted as a function of temperature in the lower panel of in Figure 3.4, for a pressure close to that found at sea level. For a temperature of 288 K, the figure shows that that $\Gamma_m = 4.67$ K km^{-1}, which is less than half as large as Γ_d.

As discussed in a later chapter, the strong sensitivity of the moist adiabatic lapse rate to temperature is important for climate change.

Atmospheric scientists love a good thunderstorm, so we want our cloudy parcel to keep rising. It will do so if it is positively buoyant, or in other words if it is warmer than the surrounding "environmental" air at the same level.[11] When conditions are right, latent heat release can make the parcel positively buoyant. This causes the parcel's initial upward displacement to become larger, which means that the hydrostatic equilibrium is unstable. We call this cumulus instability. The next time you see a cumulus cloud, remember that it was created by this simple and interesting mechanism.

The level at which the parcel becomes positively buoyant is called the "level of free convection." It turns out that the saturated parcel is positively buoyant when its moist static energy is larger than the saturation moist static energy of the environment. Figure 3.5 shows the level of free convection where the thin vertical line, representing the parcel's moist static energy, crosses the solid gray curve that represents the saturation moist static energy of the environment.

The key requirement for maintaining the rising parcel's positive buoyancy is that the upward decrease of its temperature must be slower than the upward decrease of the environmental temperature. In other words, to maintain positive buoyancy, we need the moist adiabatic lapse rate (which is followed by the parcel) to be slower than the environmental lapse rate, denoted by Γ. This is a necessary condition for cumulus instability.

A second necessary condition for cumulus instability is that the basic state must be sufficiently humid so that condensation will actually occur in a lifted parcel. Condensation can happen only if there is sufficient water vapor available. Partly for this reason,[12] cumulus instability is often called *conditional instability*.

In a basic state that is subject to conditional instability, the environmental lapse rate is steeper than the moist adiabatic lapse rate but more gradual than the dry adiabatic lapse rate. These two inequalities can be expressed mathematically by

$$\Gamma_d > \Gamma > \Gamma_m. \tag{3.9}$$

Here Γ should be interpreted as the horizontally uniform lapse rate of the idealized basic state, which, as in the case of dry static instability, is assumed to be in hydrostatic equilibrium. What we have shown is that condensation can create instability for a hydrostatically balanced basic state that would otherwise (i.e., in the absence of condensation, which is expected if the air is sufficiently dry) be gravitationally stable.

Eventually the rising parcel will lose its positive buoyancy, at a height called its "level of neutral buoyancy," where the environmental air is sufficiently warm in the sense of large dry static energy; this will happen when the parcel reaches the stratosphere, if not before, and Figure 3.5 shows the location of the level of neutral buoyancy for the given sounding. The cumulus cloud "tops out" at or slightly above its level of neutral buoyancy, and the air that ascended through the cumulus cloud tends to spread out horizontally near that level, forming what is called a stratiform cloud. The result can be a thick "anvil" cloud, similar to but often larger than the cumulus-top outflows shown in Figure 3.1. The anvil cloud eventually deteriorates into wispy cirrus clouds. Many upper-tropospheric stratiform clouds are formed by the outflows from thunderstorms. Cumulus towers occupy only a small fraction of the area, for reasons discussed below. As a result, they have only a modest effect on the flows of solar and infrared radiation, but the more extensive stratiform clouds that are produced by the outflows from the tops of the cumuli can have a big effect.

As a cumulus cloud grows, the rising cloudy air acquires kinetic energy. There is additional kinetic energy in the slowly sinking environment around the cloud and in the horizontal currents that connect the rising and sinking branches together. Where does all of this kinetic energy come from? Its source is the convective available potential energy, or CAPE. The CAPE is a measure of the "total" buoyancy of a lifted parcel, integrated over the layer across which the parcel rises. In Figure 3.5, the CAPE is proportional to the area between the thick vertical line representing the parcel's moist static energy and the dashed line representing the environmental saturation moist static energy. The region below the level of free convection contributes negatively to the CAPE and is sometimes called the "convective inhibition." The region above the level of free convection, and extending up to the neutral buoyancy level, contributes positively to the CAPE. In the example shown in the figure, the overall CAPE is positive. It turns out that the CAPE is zero when the actual lapse rate, at all levels, is the same as (or less than) the moist adiabatic lapse rate. When the lapse rate is moist adiabatic, a parcel that is rising moist adiabatically will always have the same temperature as its environment; there will be no buoyancy forces. The CAPE receives a positive contribution from layers within which the actual lapse is steeper than the moist adiabatic lapse rate, provided that the parcel is saturated. When the CAPE is positive, a parcel rising moist adiabatically can become positively buoyant.

Cumulus convection is an energy *conversion* process, in which the sum of the kinetic energy and the CAPE remains the same. Cumulus convection converts CAPE into the kinetic energy of the convection.

WIDELY SPACED TOWERS

In a weather system that contains cumulus clouds, the cumulus towers are observed to be widely spaced, with lots of clear air in between. A remarkably simple explanation of this important fact was given by Bjerknes (1938).

The starting point for Bjerknes's idea is that the environmental air has to sink to fill the space left by the rising cloudy parcels. The rising and sinking air are both warming up, but for different reasons. At a given level, the air in the updrafts is warmed, relative to the environment, by condensation. This happens because the lapse rate of the environment is steeper than the moist adiabatic lapse rate, as shown in Equation (3.9). At the same level, the environment is warmed, relative to the cloudy air, because the sinking air carries a larger dry static energy down from above. This happens because the lapse rate of the environment is less steep than the dry adiabatic lapse rate, as also shown in Equation (3.9).

To maintain mass balance, the amount of air rising in the cumulus towers has to be balanced by the amount sinking in the environment. As an example, suppose that the area of the updrafts is the same as the area of the environment (not realistic). Then mass balance is maintained when the upward speed of the cloudy parcels and the downward speed of the environment are the same. On the other hand, if the updrafts occupy much less area than the environment (realistic, and illustrated in Figure 3.6), then to maintain mass balance the cloudy

Figure 3.6. Sketch illustrating a strong updraft, represented by the large arrow, and surrounding weak downdrafts, represented by the small arrows.
The upward and downward mass flows can cancel.

parcels have to be rising faster than the environment is sinking.

Saturated rising motion is aided by positive buoyancy created through condensation, while unsaturated sinking motion must fight against the dry stable stratification. The rate of temperature increase in the updraft is proportional to the updraft speed, while the rate of temperature increase in the downdraft is proportional to the downdraft speed. Therefore, the buoyancy of the cloudy air is maximized when the updraft is fast and narrow, and the environmental sinking motion is slow and broad. This is how Bjerknes explained the observation that cumulus updrafts occupy only a small fraction of the available area.

As discussed in the Appendix to this chapter, the small fractional area occupied by the towers can be used to derive a useful formula for the vertical fluxes of energy (and other things) produced by the cumulus clouds.

WHAT DETERMINES THE INTENSITY OF THE CONVECTION?

We have established the conditions for the growth of a cumulus cloud, and what determines the cloud-top height, and why cumulus clouds are widely spaced. A very important question remains: What determines the intensity of the convection? To put it more simply, what determines how hard it rains? Here the question is *not* how hard it will rain on your house; the answer to that depends on whether a cloud happens to make a bull's-eye on your house, which means that there is a strong element of chance involved. What we are asking is what determines the rain rate as averaged over an area much larger than the cross-sectional area of a single cloud—perhaps over the area of a box 100 km on a side (10 billion m²).

As mentioned above, the kinetic energy of the convection comes from the CAPE. In order for the convection to continue, there has to be a persistent source of CAPE. There are may ways to increase the CAPE. The simplest way is to add either sensible heat or water vapor to the air near the surface, for example by surface fluxes. This increases the low-level moist static energy, and warming near the surface also tends to steepen the lapse rate. Cooling aloft also steepens the lapse rate.

Cumulus convection converts CAPE into the kinetic energy of the convection. The way this works is that the vertical profile of warming and cooling produced by the convection adjusts the environmental lapse rate toward the moist adiabatic lapse rate. Cumulus convection also dries the atmosphere by raining water onto the Earth's surface. In both of these ways, convection tends to decrease the CAPE, and so it tends to remove the instability that makes the convection possible. Just as in the case of dry convection, the convective response to instability tends to remove the instability.

Here, then, is the physical picture: Heating and moistening near the surface, and cooling aloft, can generate CAPE. Cumulus convection converts CAPE into kinetic energy, reducing the CAPE in the process. The rate at which convective kinetic energy is produced is an excellent measure of the intensity of the convection.

By adding one more simple idea, we can determine the intensity of the convection. The idea is that the convection converts the CAPE into kinetic energy just as fast as the CAPE is generated; in other words, the convection very efficiently consumes the CAPE. Under such conditions, *the rate of kinetic energy generation is equal to the rate of CAPE generation.* The rates of heating and moistening below, and the rate of cooling above, determine such measures of convective intensity such as the area-averaged rain rate.

To gain an intuitive understanding of why this is true, consider an earthy analogy. In this analogy, the convection is represented by a large, very hungry dog. The CAPE is the food in the dog's bowl. CAPE is generated when you, the dog's human companion, add food to the bowl. The ravenous dog wolfs the food down as fast as it appears, so the rate at which the dog *consumes* the food (the intensity of the convection) is equal to the rate at which you *supply* the food (the rate of CAPE generation).

Because the dog is such an efficient eater, the bowl is always nearly empty. The analogy here is that the convection consumes CAPE so efficiently that the measured CAPE is always close to zero, despite the fact that CAPE is continually generated by various processes. Because of this, the actual lapse rate is observed to be close to the moist adiabatic lapse rate (Xu and Emanuel, 1989) throughout the tropical troposphere (except in the boundary layer). For reasons explained in Chapter 4, this is true even in portions of the tropics that are far away from regions of active convection.

CUMULUS FLUXES OF ENERGY AND OTHER THINGS

Recall that the brightness temperature of the Earth as a whole corresponds to a level in the middle troposphere, where the actual temperature is close to 255 K. It follows that, in an average sense, nonradiative processes have to carry energy upward, from the surface to the middle or upper troposphere, where radiation can take over and carry the energy on out to space. Riehl and Malkus (1958) deduced from the observed energy balance and vertical structure of the tropical atmosphere that thunderstorms are the primary mechanism for upward energy transport in the tropics. Their argument is indirect; it is based on eliminating all of the other possibilities.

They began by estimating the mass circulation across a latitude 10° on the winter side of the ITCZ. They

Outflow

No flow
across
this side

Inflow

Figure 3.7. The flow through the Intertropical Convergence Zone, as analyzed by Riehl and Malkus.

neglected the mass transport across the boundary of the ITCZ on its summer side. Air flows in at the lower levels, and out at the upper levels, implying upward motion within the ITCZ, as illustrated in Figure 3.7. They then attempted to evaluate the meridional energy transports into and out of the ITCZ, as functions of height. They considered meridional transports of enthalpy, potential energy, and latent energy, which added together comprise the moist static energy. What we expect to find is a net transport of energy toward the winter pole because this is what is needed to compensate for the net radiative heating of the tropics and the net radiative cooling of the polar regions, which were discussed in Chapter 2.

The data available in 1958 were not really adequate to the task. Somehow, through inspired data analysis and perhaps a bit of luck, Riehl and Malkus arrived at results that are roughly consistent with more modern observations. They managed to get the right answer.

The low-level inflow to the ITCZ is warm and wet, while the upper level outflow is cold and dry. These simple facts make it clear that, considering the whole vertical column of air, enthalpy and latent energy flow *into* the ITCZ—the wrong way to produce poleward energy transport! The flow of potential energy can save the day. The poleward flow in the upper troposphere carries high potential energy, while the equatorward flow in the lower troposphere carries much less. The net potential energy transport is strongly out of the ITCZ.

Riehl and Malkus arrived at the estimates of moist static energy transport summarized in Table 3.1. The table shows that the moist static energy flow out of the ITCZ, at the upper levels, is slightly greater than the moist static energy flow into the ITCZ in the lower troposphere. Poleward potential energy transport does win out, as it must.

The meridional winds remove moist static energy from the ITCZ. Because energy is conserved, other processes must provide a net input of energy to the ITCZ.

Table 3.1
Net energy transports out of the ITCZ.

Layer	Moist static energy transport out of the ITCZ, in units of 10^{16} J s^{-1}
Lower troposphere	−3.67
Upper troposphere	3.80

Source: Adapted from Riehl and Malkus (1958).

From the discussion earlier in this chapter, we can deduce what must be happening: the upward turbulent flux of moist static energy from the Earth's surface into the atmosphere, which is the sum of the latent and sensible heat fluxes, must be larger than the net radiative cooling of the atmospheric column. Observations show that this is actually the case.

Table 3.1 shows that energy flows into the ITCZ in the lower troposphere, and out in the upper troposphere. Most static energy is conserved, so Riehl and Malkus concluded that there must be a net upward flow of energy inside the ITCZ. They argued, however, that this upward energy flux cannot be due to advection by the mean rising motion. If advection by the zonally averaged vertical motion were producing the upward moist static energy transport, then we would see a *large-scale* column of rising air with a vertically uniform moist static energy.

That is not observed. Figure 3.5 shows that, on the contrary, the moist static energy has a minimum at midlevels in the tropical troposphere, about 5 km above the Earth's surface. This means that large-scale vertical advection cannot be dominant. Similar reasoning shows that vertical mixing cannot explain the observed upward energy flux.

In this way, essentially through a process of elimination, Riehl and Malkus concluded, from observations, that the upward energy transport must occur in thunderstorms that penetrate through the depth of the tropical troposphere.[13] Fifty years later, their conclusion still stands.

The mechanism of the upward transport by cumulus clouds is explained in greater technical detail in the appendix to this chapter. There we show that the upward flux of moist static energy can be written as

$$F_h = M_c(h_c - \overline{h}),\tag{3.10}$$

where M_c, which is called the "convective mass flux," is a measure of the strength of the cloudy updrafts; h_c is the moist static energy in the updrafts, as represented by the thin vertical line in Figure 3.5; and $\overline{h}$ is the moist static energy averaged across the area of the entire weather system. The convective mass flux is an important measure of the vigor of the cumulus cloud system. Equations very similar to (3.10), with the same convective mass flux, can be used to express the upward fluxes of quantities other than moist static energy (e.g., the concentration of ozone).

It is far beyond our capabilities, even today, to directly measure the upward energy transport by the thunderstorms of the ITCZ. The Earth is too big, the thunderstorms are too numerous, and our observing systems are much too limited. Indirect methods of observation, like those of Riehl and Malkus, are essential in atmospheric science.

The air that rides upward in the cumulus towers carries not only energy but also water (vapor, liquid, and sometimes ice), various chemical species, and the momentum associated with the horizontal wind. The upward water transport is particularly important and interesting because, as mentioned earlier it leads to the formation of stratiform clouds in the air that has exited the cumulus updrafts. Some of the condensed water re-evaporates in the middle and upper troposphere. The net effect is that the cumulus clouds have added water vapor to the middle and upper troposphere, even though they also remove moisture from the atmospheric column as a whole by producing precipitation. The moisture that is rained out originates as the water vapor in the lower troposphere.

This long chapter has covered a lot of ground. We have focused on upward energy transport by turbulence and cumulus clouds. Both turbulence and cumulus clouds are very strongly influenced by a variety of weather systems, and they feed back to modify those systems. The next chapter gives an overview of the global circulation of the atmosphere, introduces a few types of weather systems, and explains how weather systems transport energy poleward and upward.

APPENDIX TO CHAPTER 3: MORE ABOUT EDDY FLUXES

The detailed structures and life cycles of individual small-scale eddies are not important for the global circulation;

only the collective effects of many eddies can exert significant effects on the large-scale circulation of the atmosphere. For this reason, we define vertical eddy fluxes by *averaging* over updrafts and downdrafts.

We use an overbar to represent a horizontal average at a given height and a "prime" to represent the departure from the average. For an arbitrary or generic quantity per unit mass denoted by the symbol A,[14] we write

$$A = \overline{A} + A'. \tag{A3.1}$$

This simply says that the total value of A is the sum of its average, denoted by $\overline{A}$, and the departure from that average, denoted by A'. The departure from the average, that is, the primed quantity, is interpreted as an "eddy" value. In the absence of eddies, all of the primed quantities, such as A', are zero.

We assume that the average of an average is just the same quantity back again. In other words, we assume that if a quantity has already been averaged, then averaging it again has no effect. This would mean, for example, that $\overline{(\overline{A})} = \overline{A}$. We also assume that the average of a sum is the sum of the averages. From these two assumptions, and Equation (A3.1), it follows that

$$\overline{A'} = 0, \tag{A3.2}$$

which means that the average of a primed quantity (in other words, the average of the departure from the average) is zero. Finally, we assume that

$$\overline{\overline{B}A'} = 0, \tag{A3.3}$$

where B is a second generic quantity. Equation (A3.3) means that an averaged quantity, such as $\overline{B}$, behaves like a constant.[15]

The rate at which mass is flowing up or down, per unit horizontal area, is given by the product of the density of the air, denoted by ρ, and the vertical velocity, denoted by w. The product, ρw, is called the vertical mass flux. It can be averaged, just like A; we can write

$$\rho w = \overline{\rho w} + (\rho w)' .\tag{A3.4}$$

Equation (A3.4) shows that there is an "average" vertical mass flux, given by $\overline{\rho w}$, and an eddy vertical mass flux, given by $(\rho w)'$. From the preceding discussion, we know that the average of the eddy mass flux is zero, that is, $\overline{(\rho w)'} = 0$, which means that the eddies do not produce any net vertical flow of mass. In some places mass flows up, in other places it flows down, and when we do the sums, the upward and downward mass flows cancel.

With this preparation, we can define the eddy flux of A. First, the total flux of A is defined as $\overline{\rho w A}$. Here the overbar extends over the product. Using (A3.1) and (A3.4), we write

$$\begin{aligned}
\overline{\rho w A} &= \overline{\left[\overline{\rho w} + (\rho w)'\right]\left(\overline{A} + A'\right)} \\
&= \overline{\overline{\rho w}\,\overline{A}} + \overline{\overline{\rho w}\,A'} + \overline{(\rho w)'\,\overline{A}} + \overline{(\rho w)'A'} \\
&= \overline{\rho w}\,\overline{A} + \overline{(\rho w)'A'}.
\end{aligned}\tag{A3.5}$$

On the second line of (A3.5), the middle two terms are zero because of (A3.3). The third line of Equation (A3.5) shows that the total flux of A consists of just two contributions. The first arises from the product of the average mass flux and the average value of A. This is the flux due to advection by the large-scale vertical velocity. The second is the "eddy flux," arising from the interactions of the eddy mass flux with A', which is the eddy component of A. The eddy flux is a "small-scale" advection by mass fluxes that average to zero. We adopt the notation

$$F_A \equiv \overline{(\rho w)'A'} \tag{A3.6}$$

for the eddy flux of A. The subscript indicates the quantity that is being fluxed. Equation (A3.6) says that the eddy flux of A is given by the "covariance" of the eddy mass flux and A'.

As an example, consider how Equation (A3.6) applies to the particular case of thunderstorm transports. For the updraft, we can write

$$A' = A_c - \overline{A}, \tag{A3.7}$$

where A_c is the updraft value of A. Recall that, because the environment covers almost the entire large-scale area, the value of A in the environment is very nearly the same as $\overline{A}$. It follows that for the environment $A' \cong 0$. The eddy mass flux associated with the thunderstorm updraft can be written as $(\rho w)' = \rho\sigma_c(w_c - \overline{w})$, where $\sigma_c = 1$ is the fractional area occupied by the updrafts. Recall, however, that $w_c \gg \overline{w}$. The updraft eddy mass flux, which we denote by M_c, is then approximately given by

$$M_c = \rho w_c. \tag{A3.8}$$

....

We can now write the upward flux of A due to the thunderstorms as

$$F_A \equiv \overline{(\rho w)' A'}$$
$$= \sigma_c \rho w_c (A_c - \overline{A}) - (1 - \sigma_c) \rho w'_{\text{environment}} \, A'_{\text{environment}} \qquad \text{(A3.9)}$$
$$= M_c (A_c - \overline{A}).$$

The second line of (A3.9) is just a construction, by area averaging, of $\overline{(\rho w)' A'}$. We write the primed quantities for the updraft and downdraft separately and multiply them together with area weighting. The first term is the contribution of the updrafts, and the second term, which evaluates to zero, is the contribution from the environment. The environmental contribution is negligible because $A'_{\text{environment}} \cong 0$. On the third line, we use Equation (A3.8). The quantity M_c is called the *convective mass flux*. It plays a fundamental role in determining the effects of cumulus clouds on weather systems.

There are additional "molecular" fluxes due to the random molecular motions, which give rise to molecular conduction, diffusion, and viscosity. The eddy fluxes are typically many orders of magnitude larger than the molecular fluxes, so we neglect the latter, for the most part, in atmospheric science.

It is important to remember, however, that molecular effects are ultimately responsible for the dissipation of kinetic energy, which is essentially a conversion of kinetic energy into internal energy (i.e., of macroscopic kinetic energy into microscopic kinetic energy). Similarly, the molecular thermal conductivity is ultimately responsible for the dissipation of thermal fluctuations. It is an amazing fact that even though the molecular processes act on scales of a few millimeters, they have profound effects on the global-scale circulation of the atmosphere!

The turbulent fluxes at the Earth's surface are virtually always expressed in terms of "bulk aerodynamic" formulas, for example,

$$\left(F_\theta\right)_S = \rho_S c_T |\mathbf{V}_S|(\theta_g - \theta_a), \tag{A3.10}$$

$$\left(F_q\right)_S = \rho_S c_T |\mathbf{V}_S|(q_g - q_a), \tag{A3.11}$$

$$\left(\mathbf{F}_V\right)_S = -\rho_S c_D |\mathbf{V}_S|\mathbf{V}_S. \tag{A3.12}$$

Here the subscript g denotes a value representative of the lower boundary and the subscript a represents a value representative of a level inside the atmosphere but near the surface. The quantities c_T and c_D are the heat-and-moisture transfer coefficient and the drag coefficient, respectively. In practice, the coefficients c_T and c_D are determined as empirical functions of the surface roughness and the near-surface static stability and wind shear and are used in (A3.10)–(A3.12) to compute the surface fluxes.

As can be seen from (A3.10), the surface sensible heat flux is upward when the ground (or ocean) is warmer than the air. It tends to cool the ground and warm the air, thus trying to put itself out of business. Similarly the latent heat flux is upward when the lower boundary is wetter than the air. It dries the boundary while moistening the air. Finally, as mentioned above, the surface friction transfers momentum from the air to the surface, thus tending to reduce the momentum of the near-surface air. Because all three of these surface fluxes are self-destructive, they can continue over time only if some process acts to maintain them. For example, solar radiation absorbed by the ground can maintain an upward surface sensible heat flux, subsidence drying of the air near the surface can maintain an upward surface moisture flux, and the large-scale pressure gradient can maintain a surface momentum flux.

Over a water surface, q_g is just the saturation mixing ratio evaluated using the surface temperature of the water and the surface air pressure. Over land, it is much more difficult to determine the appropriate value of q_g, which depends on such things as the soil moisture, the amount of vegetation, and the state of the vegetation (e.g., whether or not photosynthesis is occurring).

The minus sign in (A3.12) means that the near-surface momentum flux has a direction opposite to the surface wind. For example, if the near-surface wind is westerly, there will be a downward flux of westerly momentum into the Earth's surface. The ocean currents are typically very slow (centimeters per second) compared to the usual near-surface wind speeds, so for practical purposes the oceans can be considered to be at rest when air-sea momentum exchanges are considered, and that has been assumed, for simplicity, in writing (A3.12).

4 HOW ENERGY TRAVELS FROM THE TROPICS TO THE POLES

....

THE WINDS

As discussed in Chapter 2, energy has to be carried from the tropics, where solar absorption dominates, to the polar regions, where infrared loss to space wins out. It's about 10,000 km from the Equator to either pole. How does the energy make its way poleward?

The short answer is that the winds carry the energy poleward, and also upward. The winds are the motion of the air. They are associated with weather systems, which come in many shapes and sizes. There is no way that this primer can discuss the full range of weather systems, so we are going to limit ourselves to a few examples that are particularly important for poleward and upward energy transport, or particularly interesting for other reasons. The examples are the tropical Hadley cells (defined below), monsoons, midlatitude winter storms, and tropical storms.

The wind is a three-dimensional vector. As usual, there is some jargon to explain. For obvious reasons, the component of the wind vector that points in the north-south direction is particularly key to poleward energy transports. It is called the meridional wind because it blows along meridians, which are lines of constant

....

longitude, oriented north-south. By convention, the meridional wind is positive toward the north and negative toward the south. A wind from south to north is called "southerly," and a wind from north to south is called "northerly." The component of the wind vector that points in the east-west direction is called the zonal wind. By convention, a positive zonal wind blows from west to east and is called westerly. A negative zonal wind, which blows from east to west, is called easterly. The vertical wind is defined to be positive upward. You may be glad to hear that an upward wind is *not* called a "downerly"; we usually just call it rising motion. A downward wind is often called "subsidence."

On large scales, the horizontal wind can occasionally be as strong as about 100 m s^{-1}. Such strong winds are typically found in the stratosphere. The winds tend to be relatively weak near the surface because of frictional drag due to contact with the ocean or land. You know from your own experience that a typical wind speed at a height of a meter or two above the surface is just a few meters per second. The near-surface wind can be very strong in a tropical cyclone or tornado, but such storms are very small compared to the size of the Earth, and at any given location they occur very infrequently (or not at all).

On large scales, the thinness of the atmosphere ensures that vertical motions are very slow compared to the horizontal wind. A typical large-scale vertical wind speed is just a few centimeters per second, or less. As discussed in Chapter 3, in a thunderstorm, narrow vertical currents can be as strong as tens of meters per second.

THE HADLEY CELLS

Here is a very simple (too simple) idea about how the winds could transport energy poleward, an idea that might be called the "big loop theory":

Near the Equator, warm, buoyant lower-tropospheric air rises through thunderstorms to the upper troposphere. The air is blocked from entering the stratosphere because it would be negatively buoyant there. It exits the tops of the storms and begins to spread out horizontally toward both the North and South Poles. Because moist static energy is conserved following a parcel, the high moist static energy content of the air is carried poleward with it. En route to the poles, the air gradually loses energy by emitting infrared out to space. It finally sinks to the lower troposphere near the poles. Then it flows back toward the tropics, staying near the surface, to complete a gigantic loop. The low-level, equatorward branch starts out near the poles with cold, dry air that has a small moist static energy. As the air moves back toward the tropics, it is gradually heated and moistened by contact with the increasingly warm ocean and land surface. As a result, its moist static energy increases toward tropical values again. Then the cycle repeats.

The big loop theory is wrong, but it can serve as a useful starting point for an analysis of atmospheric energy transport. Let's immediately turn to observations of the atmospheric circulation to see whether or not there is any empirical support for the big loop theory. The observations come from a huge global observing network,

supported by many nations, that includes surface measurements, data gathered aloft from weather balloons
and aircraft, and a wide variety of instruments on
weather satellites.

Figure 4.1 shows the observed pattern of meridional
and vertical motion, averaged around latitude circles.
We call it the "mean meridional circulation," or MMC.
The upper panel shows observations for January, and the
lower panel for July. The data are plotted as functions
of latitude and height. The contours in the plots show
what is called the "streamfunction" of the MMC. When
the contours of the streamfunction are vertical, the air is
moving up or down. When the contours are horizontal,
the air is moving horizontally. The plot thus gives a simple, intuitive picture of how the meridional and vertical
winds carry mass around. The units of the streamfunction are 10^{12} g s^{-1}. This unit is sometimes called a "Sver-
drup," especially in the oceanographic literature.

The most conspicuous features in Figure 4.1 are
roughly circular "cells" in the tropics. The positive cells
carry mass counterclockwise in the plots, and the negative cells carry mass clockwise. In January, there is a big
positive cell between about 10°S and 30°N. The upper
branch of the cell carries mass toward the North Pole,
which is the winter pole. The lower branch carries mass
back toward, and eventually across, the Equator. In July
there is a big negative cell between about 15°N and 30°S,
in which the upper branch again carries mass toward
the winter pole, this time to the south, and the lower
branch again carries mass back toward the Equator.

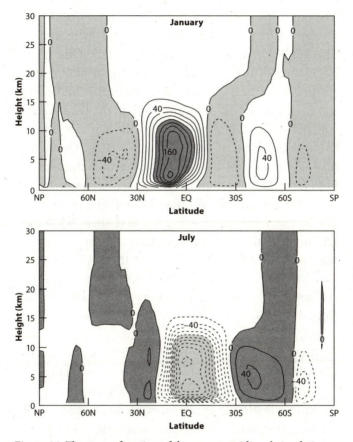

Figure 4.1. The streamfunction of the mean meridional circulation, plotted as a function of height and latitude.

Positive values, denoted by solid contours, represent counter-clockwise circulations, while negative values, with dashed contours, represent clockwise circulations. Strong positive values are darkly shaded, and strong negative values are lightly shaded. The units are 10^{12} g s^{-1}.

The big tropical cells are called the "Hadley cells," after George Hadley, an 18th-century English meteorologist. The Hadley cells are major components of the global circulation of the atmosphere. They contribute most of the poleward atmospheric energy transport in the tropics. Each Hadley cell resides mostly in the winter hemisphere but extends slightly across the Equator into the summer hemisphere. As the seasons change, the meridional component of the wind reverses over the Equator, both near the surface and in the upper troposphere.

The large Hadley cells have their rising branches in the summer hemisphere tropics and their bodies extending into the winter hemisphere subtropics. The rising branches mark the location of the ITCZ discussed in Chapter 3. The sinking branches are found in the subtropics. The sinking air is very dry, and this is why the subtropics are the home to the major deserts of the world, such as the Sahara.

The strengths of their mass circulations, in terms of the streamfunction, are about 160 Sverdrups (Figure 4.1).

The Hadley cells have their rising branches in warmer air and their sinking branches in cooler air. They are essentially buoyancy-driven and obtain their kinetic energy by conversion from gravitational potential energy. The latent heat release in thunderstorms makes it possible for the air in the rising branch of a Hadley cell to rise from near the surface to near the tropopause. Similarly, radiative cooling by infrared emission makes it possible for the air in the sinking branch of a Hadley cell to descend from the tropopause to the Earth's surface.

There are additional features in the plots, besides the big Hadley cells. First of all, there is a smaller Hadley cell in both seasons, on the opposite side of the Equator from the big Hadley cell. Figure 4.1 also shows additional, weaker circulations in the middle and high latitudes, most clearly in the Southern Hemisphere in both seasons and both hemispheres. These are called "Ferrel cells." The sinking branches of the Ferrel cells coincide with the sinking branches of the Hadley cells; both are found in the subtropics, near 30° north and south of the Equator. The rising branches of the Ferrel cells are cooler than the sinking branches, so buoyancy actually fights against them. They are driven by the mechanical energy of midlatitude storms. Finally, the polar regions play host to weak circulations with (relatively) warm rising branches and cooler sinking branches.

The alternating bands seen in Figure 4.1 are somewhat reminiscent of the banded structures seen in the atmospheres of Jupiter and the other gas giant planets of our Solar System. This is not a coincidence. Banded structures are characteristic of the atmospheric circulations of rapidly rotating planets. The number of bands tends to increase as the rotation rate increases (as the length of day decreases), all else being equal.

The Hadley cells carry air with large moist static energy poleward and air with smaller moist static energy back toward the Equator. This was foreshadowed in Chapter 3, in connection with the article of Riehl and Malkus. The Hadley cells are, therefore, somewhat similar to the hypothetical "big loop" circulations described at the beginning of this chapter, except that the big loops were imagined to stretch from the Equator to the poles, while the Hadley cells extend to only about 30° of latitude in each hemisphere. An explanation is given later.

The air circulating in a Hadley cell crosses more than 30° of latitude. Thanks to the Earth's rotation, this has important implications for the zonal wind. The explanation involves angular momentum. Recall from physics that the angular momentum is a vector, but because the Earth is rapidly rotating, one particular component of the atmosphere's angular momentum is much larger than the others. It is the component that points in the direction of the Earth's axis of rotation (i.e., toward the North Star). This component is denoted by the symbol M, and its magnitude per unit mass can be written as

$$M \equiv (\Omega a \cos \varphi + u) a \cos \varphi. \tag{4.1}$$

Here Ω is the magnitude of the angular velocity of the Earth's rotation (2π divided by the length of a sidereal day,[1] which is 86,146 s), a is the radius of the Earth (about 6,400 km), u is the zonal wind, and φ is latitude. You can understand where (4.1) comes from by parsing it out term by term. The distance from the Earth's axis of rotation to a point on the Earth's surface at latitude φ is given by $a \cos \varphi$. The quantity $\Omega a \cos \varphi$ is the speed with which a point attached to the spinning Earth is moving toward the east (Figure 4.2). The sum $\Omega a \cos \varphi + u$ is the total speed toward the east of an air parcel that is moving toward the east, relative to the spinning Earth, with zonal

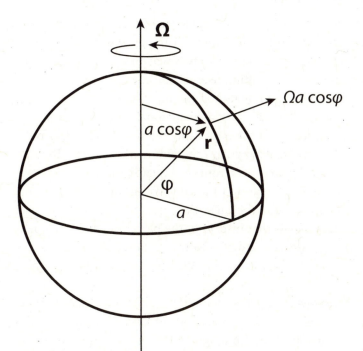

Figure 4.2. Sketch illustrating the rotating Earth.

The Earth's rotation is indicated by the vector (boldface) pointing upward, that is, toward the North Star. The position of a point on the Earth is indicated by the position vector **r**. The point is moving toward the east at the speed $\Omega a \cos \varphi$.

wind u. Finally, this total speed toward the east is multiplied by the moment arm $a \cos \varphi$, to give the angular momentum per unit mass.[2]

The angular momentum of an air parcel can be changed only by the torques associated with pressure forces and friction. Except for the effects of those torques, the angular momentum of an air parcel is conserved, and this conservation property is what makes the concept of angular momentum useful. Consider a parcel that starts out near the Equator, at the top of the rising branch of the Hadley cell, with a small zonal wind, that is, $u \cong 0$. Suppose that the parcel moves toward the winter pole, conserving its angular momentum as it travels. The further the parcel gets from the Equator, the smaller $a \cos \varphi$ becomes. Equation (4.1) says that, in order for M to remain constant (because it's conserved), u has to increase. This means that instead of the parcel moving due north or south (toward the North or South Pole), it curves around toward the east. We say that the wind becomes "westerly."

How strong is the westerly wind? Recall that the Hadley cells extend to about 30° of latitude away from the Equator. Using Equation (4.1), you can figure out how fast the parcel would be moving toward the east when it reaches that latitude, assuming that it started out with $u \cong 0$ on the Equator. The answer turns out to be about 140 m s^{-1}. That's a pretty strong wind! It is actually too strong because in reality pressure forces and friction reduce the angular momentum as the parcel travels, but our little "back of the envelope" calculation is giving at least the right general idea, as you can see by looking at Figure 4.3, which shows how the zonal wind varies with latitude and height, in January and July. There are strong westerly winds at about 30° away from the Equator

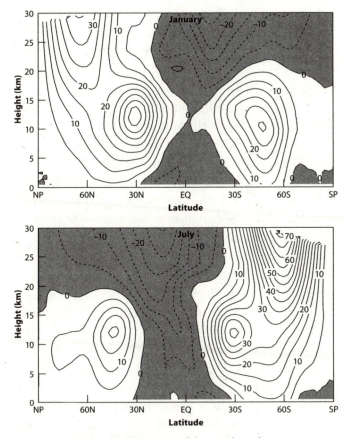

Figure 4.3. Latitude-height section of the zonal wind.
The contour interval is 5 m s^{-1}. Easterlies are shaded.

and about 12 km above the surface—just about where the Hadley cells stop. These westerlies are called the jet streams. Here we are, just a few pages into Chapter 4, and you already have some idea why there are jet streams. They are important elements of the Earth's climate.

Our parcel gets caught up in the subsiding (sinking) branch of the Hadley cell. As it sinks toward the surface, friction slows down its westerly motion, and it arrives near the surface with a small westerly zonal wind. The lower-tropospheric branch of the Hadley cell then carries it back toward the Equator. Because the parcel's latitude is changing, angular momentum conservation comes into play again. Friction in the boundary layer limits the speed of the parcel to just a few meters per second relative to the rotating Earth, but a tendency to conserve angular momentum implies, via Equation (4.1), that the parcel will turn toward the west. In other words, the winds become easterly. And yes, Figure 4.3 does show that the tropical low-level winds are easterly.

In centuries past, people relied on the maritime winds to drive sailing ships. You may remember, from a history or geography class, that the tropical "trade winds" blow from east to west and facilitate shipping routes toward the west. Now you know why the trade winds are there. They too are important elements of the Earth's climate.

The jet streams are closely related to winter storms in the middle latitudes. We will discuss those connections a little later in this chapter, but before leaving the tropics, let's talk about monsoons.

MONSOONS

A monsoon can be thought of as a locally intensified Hadley cell. The intensification is local in the sense that it is confined to a particular range of longitudes. It is brought about by the temperature contrast between an ocean and a continent, often enhanced by the effects of mountains. The air over a summer continent is strongly warmed by the Sun, while the air over the adjacent ocean remains relatively cool because of the ocean's much larger heat capacity. This temperature difference between land and sea drives onshore winds carrying moisture that has recently been evaporated from the sea. Thunderstorms form, making it even easier for the air to rise over the warm land.

The most powerful monsoons on Earth occur in the vicinity of southern Asia and Australia. In the Northern Hemisphere summer, and in that range of longitudes, the rising branch of the Hadley cell is well north of the Equator, near the southern edge of the Himalayan mountains. Air that originates over the Southern Hemisphere ocean flows westward toward Africa, then north across the Equator, near the African east coast. All along the way it accumulates water vapor, evaporated from the warm tropical Indian Ocean. It finally flows swiftly onto the Asian continent from the south and west, making landfall on the west coast of India. Tremendous rains fall over the Indian subcontinent and southeast Asia, and also over the adjacent Indian Ocean. Meanwhile, strong sinking motion occurs in the adjacent portions of the Southern Hemisphere, including over Australia.

....

In January, on the other hand, the air over India and southeast Asia is sinking. It streams southward off of the cold Asian continent, crosses the Equator, and flows onto Australia from the north and west, carrying water vapor evaporated from the Indian Ocean. The rising branch of the Hadley cell occurs over northern Australia, where spectacular rain events can occur.

This seasonal reversal of the low-level winds is characteristic of monsoons, and in fact the original meaning of the word "monsoon" relates to seasonally reversing winds. Recall that seasonally reversing meridional winds are also characteristic of the Hadley cells.

Smaller monsoons also occur in several other parts of the world, including western Africa, the Amazon basin of South America, and even southwestern North America. In each case, moist air flows onto a warm summer continent from the adjacent ocean. The rain provided by monsoons is essential to agriculture and other human endeavors throughout the tropics.

HOW THE EARTH'S ROTATION AFFECTS THE WIND, PRESSURE, AND TEMPERATURE

Maps of the "sea level pressure" are shown in Figure 4.4. The maps show climatological averages for the months of January and July. The sea level pressure is actually the surface pressure, with an adjustment to remove the strong features of the surface pressure that are associated with mountains. For practical purposes, you can think of the sea level pressure map as depicting the horizontal

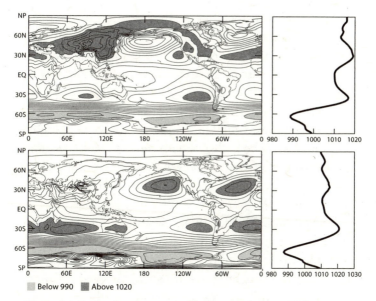

Below 990 Above 1020

Figure 4.4. Sea level pressure maps for January (top) and July (bottom).

The units are hundreds of pascals, abbreviated hPa. The contour interval is 3 hPa. Values higher than 1,020 hPa have dark shading, and those lower than 990 hPa have light shading. The zonal averages are shown on the right.

pressure differences that affect the winds near the surface. There are many interesting features in the maps of Figure 4.4. Most of them have conventional names, which are known and used mainly by atmospheric scientists. For example, the large low-pressure center over the North Pacific Ocean in January is called the Aleutian Low because it is found near the Aleutian Island chain, southwest of Alaska. The sea level pressure contours are packed close together in the Aleutian low, which means that there is a strong pressure difference between the center of the low and the surrounding regions. The large high-pressure center in the north Atlantic in July is called the Bermuda High.

Based on everyday experience and simple physical reasoning, the air would be expected to flow out of highs and into lows. The high-pressure centers would become weaker as the air flowed away from them, and the low-pressure centers would become weaker as the air flowed into them.

As you probably already know, this is *not* what typically happens in the global circulation of the atmosphere. Instead, the air circulates counterclockwise around Northern Hemisphere lows, in what are called cyclonic circulations, and clockwise around Northern Hemisphere highs, in anticyclonic circulations. In the Southern Hemisphere, cyclonic circulations (around lows) are clockwise and anticyclonic circulations (around highs) are counterclockwise. Subtropical highs are found off the west coasts of North America, South America, Australia, and southern Africa and over the northeastern Atlantic in the northern summer. These are all regions of cold ocean temperatures. The low-level winds of the subtropical highs blow equatorward along the west coasts of the continents, and then turn westward to join the trades. The connections between subtropical highs and sea surface temperatures are discussed in Chapter 8.

Cyclonic and anticyclonic circulations are made possible by the rotation of the Earth. Suppose that an air parcel is released just outside a Northern Hemisphere low, with no velocity relative to the Earth's surface. It feels a pressure difference that pushes it toward the center of the low. As soon as it starts to move toward lower pressure, however, it is deflected toward the right of its direction of motion. If it is moving north, the rightward deflection is toward the east, if it is moving west, the deflection is toward the north, and so on. If the parcel moves in such a way that the rightward deflection always points away from the low-pressure center, then it will circle counterclockwise around the low. At each point along its (roughly) circular path, the pressure force pushes it toward the low, and the rightward deflection pushes it away from the low.

Why does this happen? The parcel deflections described above are due to the Earth's rotation, through what is called the Coriolis effect. The deflections are not the result of a true force; in fact, they are not really deflections at all. Both the force and the deflections are "apparent." The best way to understand the simple nature of the Coriolis effect is to watch what happens when two people toss a ball back and forth while riding on a merry-go-round. I can't embed a movie into this book, but I strongly recommend that you do an Internet search for videos associated with the word "Coriolis." You will find *many* video demonstrations of the Coriolis effect based on games of merry-go-round catch. You can reproduce the demonstrations with a ball and a friend at a neighborhood park.

What the videos show is very simple. Newton's first law of motion says that a moving parcel will continue in a straight line at constant speed unless it is acted upon by a force. If you toss a ball to your friend, while you are both riding on the merry-go-round, the ball actually does travel in a straight line at constant speed, just as Newton said. The ball's straight-line motion is readily apparent to a bystander who is not riding on the merry-go-round. But because you and your friend are rotating with the merry-go-round, it *appears* to you both that the ball is deflected to the right if the merry-go-round is turning counterclockwise (as seen from above), and to the left if the merry-go-round turns clockwise. The ball *appears* to fly away from the merry-go-round, as if it had been pushed. This is the (apparent) deflection due to the Coriolis effect.

Now you see why the Coriolis effect is called an "effect" rather than a force. It is not a force. It is sometimes called an apparent force.

The sign of the Coriolis effect changes across the Equator. To see why, consider a case in which the merry-go-round is rotating counterclockwise *as seen from above*. We will (arbitrarily) stipulate that an observer looking from above is in the Northern Hemisphere. A person in the Southern Hemisphere, being "upside-down,"[4] is looking up on the merry-go-round from below. From the point of view of the upside-down observer, the merry-go-round is rotating clockwise. Seen from above (north), the thrown ball is deflected to the right. Seen from below (south), it is deflected to the left. To both observers (above and below, north and south), the ball appears to fly away from the merry-go-round.

....

The magnitude of the Coriolis effect can be measured by the Coriolis parameter,[37] which is given by $f \equiv 2\Omega \sin\varphi$. The Coriolis parameter is positive in the Northern Hemisphere, zero on the Equator, and negative in the Southern Hemisphere.

Let's return now to the low-pressure center discussed above, in which the pressure force is pushing parcels toward the low and the rightward deflection due to the Coriolis effect is pushing them away from the low. In terms of our merry-go-round analogy, the pressure force is a true force that pushes the ball toward the center of the merry-go-round (toward low pressure), preventing it from "flying away," which is what it would appear to do if no force were acting on it. The pressure force, being a real force, causes the ball to accelerate, that is, to deviate from straight-line motion at constant speed. The ball accelerates toward the center of the merry-go-round. The apparent force due to rotation (i.e., the Coriolis effect) is balanced, from the perspective of a rotating observer, by the true force associated with the pressure gradient.

The motion of the air is governed by—wait for it—"the equation of motion," which is an expression of Isaac Newton's Second Law of Motion. For those who are interested, a fairly exact version of the equation of motion is derived in the appendix to this chapter. Here we will just use an approximate form, called *geostrophic balance*. The approximation is accurate away from the Equator, where the Coriolis effect is strong. In geostrophic balance, the pressure force and the Coriolis effect are equal and opposite. Geostrophic balance can be expressed mathematically by

$$-fv = -\frac{1}{\rho}\left(\frac{\partial p}{\partial x}\right), \text{ and } fu = -\frac{1}{\rho}\left(\frac{\partial p}{\partial y}\right). \tag{4.2}$$

Here u is the component of the horizontal wind that points toward the east and v is the component that points toward the north. The left-hand sides of these two equations represent the components of the Coriolis effect, and the right-hand sides represent the components of the pressure-gradient force. The partial derivatives of pressure in (4.2) measure how rapidly the pressure is changing with distance x toward the east and y toward the north. Observations show that the large-scale winds in the Earth's atmosphere are nearly in geostrophic balance, except close to the Equator.

When the wind is in geostrophic balance, parcels move along or parallel to lines of constant pressure, rather than across them. Therefore, in geostrophic balance, the pressure force acts at right angles to the direction of the wind. Recall from elementary physics that the rate of work done by a force is the magnitude of the force times the rate of motion *in the direction of the force*. In geostrophic balance, the direction of the pressure force is perpendicular to the direction of the wind, so it does no work on the air and has no effect on the kinetic energy of the air. For this reason, an idealized geostrophically balanced circulation can continue indefinitely, without an energy source.

Close to the Equator, where the Coriolis parameter is small, the effects of rotation are weak and geostrophic balance is less prevalent. The air near the Equator responds to pressure differences in much the same, familiar way as the air in the heating and air conditioning ducts of a house. When a pressure difference exists in the duct system, the air flows from high to low pressure. If nothing acts to maintain the pressure difference, high pressure "drains down," the low pressure "fills up," and the pressure difference goes away. A pressure difference can be maintained over time only by a powered fan or other mechanism that does work on the air. Turn the fan off, and the pressure quickly becomes uniform throughout the system.

Look at Figure 4.4 again. Notice that the isolines of pressure are far apart in the tropics. This means that the tropical pressure differences are weak. The explanation is that the Coriolis effect is weak in the tropics, and air moves quickly to drain highs and fill lows, just as in the duct system described above. In middle and high latitudes, where the Coriolis effect is strong, much larger horizontal pressure differences can be maintained, and are observed, as shown in Figure 4.4.

For the reasons given above, the pressure is (approximately) horizontally uniform throughout the tropics, and this is true *at every height*. It follows that the temperature must also be horizontally uniform at every height. To see why, consider the hydrostatic balance discussed in Chapter 1, rewritten, using the ideal gas law, as

$$\frac{\partial p}{\partial z} = -\frac{pg}{RT}. \qquad (4.3)$$

If the pressure is horizontally uniform at every height, then a horizontal derivative of the left-hand side of (4.3) must be zero. It follows that a horizontal derivative of the right-hand side must also be zero. That can happen only if the temperature is horizontally uniform. We conclude that *the temperature must be (almost) horizontally uniform throughout the tropics* because the pressure is (almost) horizontally uniform there. The larger horizontal variations of temperature and pressure in midlatitudes are made possible by the stronger midlatitude Coriolis effect.

As explained in Chapter 3, the actual lapse rate in regions of tropical convection is observed to be close to the moist adiabatic lapse rate (except near the surface). This is true because the cumulus clouds consume convective available potential energy as fast as it is made available, and the temperature inside the cumulus clouds themselves follows the moist adiabatic lapse rate.

Now connect that idea with our newest result, which is that the temperature is horizontally uniform throughout the tropics (except near the surface). Together, these two facts imply that the temperature follows the moist adiabatic sounding throughout the tropics (except near the surface), even far away from regions of deep convection. We are thus led to an amazing and very important conclusion: tropical thunderstorms, which, as Bjerknes explained, occupy only a tiny fraction of the area of the tropics, are able to set the temperature sounding throughout the tropical troposphere, even far away from the convectively active regions! Thunderstorms are the 800-pound gorillas of the tropical atmosphere.

THE THERMAL WIND

Especially during winter, the midlatitude temperature can change very rapidly in the horizontal (and in time). It follows from Equation (4.3) that, when the temperature changes rapidly in the horizontal, the horizontal pressure differences change rapidly with height. Since geostrophic balance says that the strength of the wind is proportional to the horizontal pressure differences, *the wind changes rapidly with height when the temperature changes rapidly in the horizontal.* This is called the "thermal wind" relationship. In particular, it turns out that the zonal (west to east) wind increases upward when the temperature decreases toward the poles. This can happen only away from the Equator, where the temperature can change rapidly in the horizontal. Here is some more jargon: when the temperature changes rapidly in the horizontal and the wind changes strongly with height, the atmosphere is said to have a "baroclinic" structure.

Now look back at Figure 4.3, which shows how the zonal wind varies with latitude and height. At 30°N in January, the zonal wind increases strongly upward from the surface to the level of the jet stream. The thermal wind relationship discussed above implies that, below the level of the jet stream, the temperature decreases rapidly toward the North Pole. So, is it true? Figure 4.5 shows that it is. The figure shows the zonally averaged temperature as a function of latitude and height, for January and July. The isolines of temperature tilt strongly

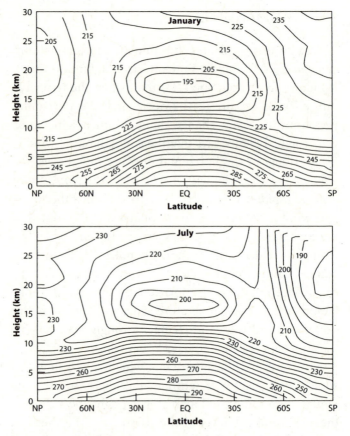

Figure 4.5. Latitude-height cross sections of temperature, for January and July.

The contour interval is 5 K.

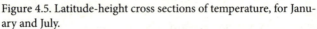

downward toward the north near 30°N in January, especially in the lower troposphere. This downward tilt means that the temperature is decreasing rapidly toward the north. We say that the midlatitude troposphere has "baroclinicity." The summer midlatitudes are also baroclinic, but less so.

Notice also that the isolines of temperature are very flat in the tropics. This confirms the horizontal homogeneity of tropical temperature, discussed earlier in this chapter.

As seen on the right-hand side of Figure 4.4, the zonally averaged sea level pressure varies strongly with latitude. The highest zonally averaged sea level pressures are found in the subtropics, where the world's major deserts occur. Smaller pressures are found near the Equator and in middle latitudes. The corresponding near-surface winds are easterly (blow toward the west) in the tropics and westerly (blow toward the east) in middle latitudes. This is consistent with the pattern of winds shown in Figure 4.3.

WINTER STORMS

I'm revising this chapter in early February 2011. A tremendous winter storm has just blasted the eastern two-thirds of North America. The northeast was visited by snow and ice, high winds, and very cold temperatures. Figure 4.6 shows an example of the type of weather system responsible, which frequently occurs in the middle latitudes, especially in winter. These beautiful storms are

Figure 4.6. A winter storm over North America on October 26, 2010. The core of the storm is in the upper Midwest, near the top center of the image. South and east of the center, warm, humid air is flowing poleward. North and west of the center, cool and relatively dry air is flowing equatorward.

Source: http://www.crh.noaa.gov/images/dlh/StormSummaries/ 2010/october26/satellite_large_2132z.JPG.

very important for transferring energy poleward, especially in the winter hemisphere. How do they work?

The strong poleward decrease of temperature in the midlatitude winter, seen in Figure 4.5, is associated with *baroclinic instability*, which is responsible for the growth of winter storms. These storms, which are also called "baroclinic waves," contain cold fronts and warm fronts, which are familiar to anyone who lives in middle latitudes. The distinction between cold and warm fronts is based entirely on the direction of the front's motion with respect to the Earth's surface. A cold front moves in such a way that warm air is replaced by cold air at a given location; a warm front moves so that cold air is replaced by warm air. A front of either variety is characterized by rapid, nearly discontinuous spatial changes in temperature and wind. On the west side of a front, the air is cold and the wind is generally blowing from the direction of the pole, for example, from the north in the Northern Hemisphere. On the east side of a front, the air is warmer and the wind is generally blowing from the direction of the Equator.

Recall that instability is often discussed in terms of small perturbations of a "basic state" that is in equilibrium. In the case of baroclinic instability, the basic state is zonally uniform, statically stable, and in geostrophic (and hydrostatic) balance. It is also baroclinic, in that the temperature decreases toward the pole and the westerly zonal wind increases with height. The idealized basic state has no fronts.

When conditions are right for baroclinic instability, a small disturbance that is added to this basic state will amplify. In a growing perturbation, the cold high-latitude air slides under the warmer low-latitude air. The cold air on the poleward side moves downward and toward the Equator, making a cold front, and the warm air on the tropical side moves upward and toward the pole, making a warm front. Because the heavy cold air sinks and the light warm air rises, thermal energy is transported upward, and at the same time the center of mass of the air column descends, releasing gravitational potential energy. Because the cold air moves toward the tropics and the warm air moves toward the poles, baroclinic waves contribute to the poleward energy transport by the atmospheric circulation.

The rising warm air from near the surface carries water vapor with it. Cloud formation follows, as can be seen in Figure 4.6. The release of latent heat makes it easier for the warm air to rise. Baroclinic waves sometimes produce heavy precipitation, especially in winter, and as a result the zonally averaged precipitation rate has maximums in middle latitudes of both hemispheres, as will be shown in Chapter 6.

Baroclinic waves occur in both winter and summer, but because the temperature difference between the tropics and the pole is much larger in the winter hemisphere, the storms are much more intense in winter. When summer comes, they weaken and shift closer to the pole.

It is useful to compare and contrast baroclinic instability with the convective instability discussed in Chapter 3. There are both similarities and differences. An important similarity is that both are associated with the conversion of gravitational potential energy into kinetic energy. An important difference is that the convective instability gives rise to thunderstorms, which are typically just a few kilometers wide, while baroclinic instability gives rise to storms that are thousands of kilometers across, like those shown in Figure 4.5. Recall that convective instability is sensitive to the rate at which temperature changes with height and requires a basic-state vertical temperature profile that is hydrostatically unstable. In contrast, baroclinic instability is sensitive to the rate at which temperature changes in the horizontal. In the basic state for baroclinic instability, the lapse rate can be gravitationally stable with respect to both dry and moist convection. The low-level temperature decreases strongly toward the winter pole, and the zonal wind increases strongly upward. Cumulus instability can occur only when clouds form. Baroclinic instability can be affected by the formation of clouds, but (in principle) it can occur in the absence of clouds.

What is the relationship between the Hadley cells and the baroclinic waves? The Hadley cells do not penetrate into middle latitudes, and baroclinic waves do not penetrate into the tropics. The two types of weather systems interact with each other in the subtropics. One point of view is that the Hadley cells are limited to the tropics because the pattern of winds and temperatures associated with a Hadley cell becomes baroclinically unstable in middle latitudes. The instability prevents the Hadley cells from penetrating further poleward and gives rise to the baroclinic waves. The energy that is carried to the subtropics by the Hadley cells is "handed off" to the baroclinic waves, which carry it on toward the poles.

TROPICAL STORMS

Large-scale winter storms are important, but what about tropical storms? "Tropical cyclone" is a generic term for what are called "hurricanes" or "typhoons," depending on where they occur.[5] Tropical cyclones are spectacular and very destructive.

As explained in the excellent book by Kerry Emanuel (see the suggestions for further reading at the end of the book), tropical cyclones are produced by the fluxes of energy into the atmosphere from the warm tropical oceans. The strong winds of such a storm increase the rate at which latent heat is supplied to the atmosphere by the ocean, and the increased energy supply promotes an intensification of the tropical cyclone. This self-perpetuating process leads to an amplification of the storm, with maximum wind speeds near the surface exceeding 70 m s^{-1} in extreme cases. Tropical cyclogenesis is very different from the baroclinic instability that leads to the growth of storms in middle latitudes.

Tropical storms originate in the tropics, but they can and often do move into midlatitudes. They weaken when they move over cool water, and they weaken very rapidly when they move over land. Even a weakened tropical storm can produce tremendous rainfall, however.

What is the role of tropical cyclones in climate? First of all, the existence of such storms is in itself an important aspect of the climate. If the annual number or maximum intensity of tropical storms changed significantly, that would be an important aspect of climate change.

The time interval between tropical storms is fairly long, however. On the average, even a storm-prone location may experience a strong tropical cyclone only once every few years. In contrast, strong winter storms visit midlatitude cities like New York at least several times each winter.

Despite their relatively infrequent visits to any given location, tropical cyclones produce so much rain that they contribute a significant fraction of the climatological summer precipitation in such places as eastern North America. They cool the oceans by evaporating water from the surface and by mixing cooler water up from below. The energy that they extract from the oceans is carried into the upper troposphere, ultimately helping to balance the Earth's energy budget by allowing energy to leak back out to space.

SUMMARY, AND WHAT'S NEXT

Chapters 2–4 have presented a brief overview of the most important mechanisms of atmospheric energy flow. At this point, a summary is in order because the remaining chapters of this book deal with other things.

Satellite data show that the Earth absorbs more energy than it emits in the tropics, and emits more than it absorbs at higher latitudes. A poleward flow of energy is required to maintain balance.

To a first approximation, the energy from the Sun passes through the relatively nonabsorbing atmosphere and is absorbed by the Earth's surface. A lot of additional energy is added to the surface in the form of infrared radiation emitted downward by the atmosphere. The surface cools in part by emitting infrared radiation. The net radiative energy absorbed by the surface is primarily used to evaporate water.

An upward flow of energy is required to balance the energy input to the atmosphere from the surface and to permit the energy to escape to space. The upward energy transport is accomplished by infrared radiative transfer through the absorbing and emitting atmosphere, by turbulence near the surface, by thunderstorms, and by baroclinic waves in midlatitudes.

The latent heat of the water vapor is released when clouds form and precipitate. Latent heat release makes it possible for the air to rise in the ascending branch of a Hadley cell, and the poleward and equatorward branches of the Hadley cells transport energy from the tropics into the subtropics. Midlatitude storms, associated with baroclinic instability, carry the energy from the subtropics to the polar regions.

A climate change involves perturbations, large or small, to all of the various processes involved in atmospheric energy transport. The surface temperature, the infrared opacity of the atmosphere, precipitation and evaporation, the Hadley cells, and baroclinic waves are all affected by externally forced changes in the climate system. And this brings us to the subject of atmospheric feedbacks on climate change.

APPENDIX TO CHAPTER 4: CONSERVATION OF MOMENTUM ON A ROTATING SPHERE

Newton's equation of motion, as applied in an inertial (i.e., nonaccelerating) frame of reference, can be written as follows:

$$\frac{D_a \mathbf{V}_a}{Dt} = -g\mathbf{e}_r - \alpha \nabla p - \mathbf{F}. \qquad (A4.1)$$

Here $\frac{D_a(\ )}{Dt}$ is the Lagrangian derivative in the inertial frame, and $\mathbf{V}_a$ is the velocity as seen in the inertial frame. We use the boldface, nonitalicized font to represent vectors. The left-hand side of (A4.1) represents the acceleration of the air as seen in an inertial frame. The precise meaning of the acceleration $\frac{D_a\mathbf{V}_a}{Dt}$ is clarified below. The effects of gravity are represented by $-g\mathbf{e}_r$, where g measures the strength of the Earth's gravity and $\mathbf{e}_r$ is a unit vector pointing outward away from the center of the Earth. The last term of (A4.1), $-\mathbf{F}$, represents the effects of friction.

The physical meaning of (A4.1) is that the acceleration of a parcel of air, written on the left-hand side, is due to the combined effects of the three forces per unit mass listed on the right-hand side. They are gravity, the pressure-gradient force, and friction.

Now consider a reference frame that is rotating with the Earth. This is the reference frame that we perceive as we live our lives on our spinning planet. A rotating coordinate system is not an inertial frame, so Equation (A4.4) must be transformed to describe momentum conservation in the rotating frame.

The length of a sidereal day is 86,164 s, so the Earth rotates about its axis with an angular velocity of $\frac{2\pi}{(86,400\text{ s})} = 7.292 \times 10^{-5}$ s^{-1}. This angular velocity can be represented by a vector, Ω, pointing toward the celestial North Pole, as shown in Figure 4.2. Let $\mathbf{r}$ be a position vector extending from the center of the Earth to a parcel of air whose position is generally changing with time. The "absolute velocity" of the air, $\mathbf{V}_a \equiv \frac{D_a\mathbf{r}}{Dt}$, is related to its relative velocity, $\mathbf{V} \equiv \frac{D\mathbf{r}}{Dt}$, as seen in the rotating coordinate system, by

$$\frac{D_a \mathbf{r}}{Dt} = \frac{D\mathbf{r}}{Dt} + \mathbf{\Omega} \times \mathbf{r}, \tag{A4.2}$$

or

$$\mathbf{V}_a = \mathbf{V} + \mathbf{\Omega} \times \mathbf{r}. \tag{A4.3}$$

Because in general the air is moving with respect to the Earth, $\mathbf{r}$ is changing with time as seen in the rotating frame. According to (A4.3), the velocity as seen in the rotating frame is different from the velocity as seen in the inertial frame.

A transformation of the form (A4.2) can be applied to any vector; above, we applied it to the position vector, $\mathbf{r}$. Now we apply it to the acceleration vector. The acceleration in the inertial frame is related to the acceleration in the rotating frame by

$$\frac{D_a \mathbf{V}_a}{Dt} = \frac{D\mathbf{V}_a}{Dt} + \mathbf{\Omega} \times \mathbf{V}_a. \tag{A4.4}$$

According to (A4.4), the acceleration in the rotating frame is different from the acceleration in the inertial frame. Substituting for $\mathbf{V}_a$, from (A4.3), we find that

$$\begin{aligned}
\frac{D_a \mathbf{V}_a}{Dt} &= \frac{D}{Dt}\left(\mathbf{V} + \mathbf{\Omega} \times \mathbf{r}\right) + \mathbf{\Omega} \times \left(\mathbf{V} + \mathbf{\Omega} \times \mathbf{r}\right) \\
&= \frac{D\mathbf{V}}{Dt} + 2\mathbf{\Omega} \times \mathbf{V} + \mathbf{\Omega} \times \left(\mathbf{\Omega} \times \mathbf{r}\right).
\end{aligned} \tag{A4.5}$$

This relates the absolute acceleration, $\frac{D_a \mathbf{V}_a}{Dt}$, to the apparent acceleration, $\frac{D\mathbf{V}}{Dt}$, as seen in the rotating frame.

Using (A4.5) in (A4.1), we can now write the equation of motion relative to the rotating frame as

$$\frac{D\mathbf{V}}{Dt} = -2\mathbf{\Omega} \times \mathbf{V} - \mathbf{\Omega} \times (\mathbf{\Omega} \times \mathbf{r}) - g\mathbf{e}_r - \alpha \nabla p - \mathbf{F}. \qquad \text{(A4.6)}$$

The term $-2\mathbf{\Omega} \times \mathbf{V}$ represents the Coriolis acceleration, whose direction is perpendicular to $\mathbf{V}$. The term $-\mathbf{\Omega} \times (\mathbf{\Omega} \times \mathbf{r})$ represents the centrifugal acceleration. It is very small compared to the other terms of (A4.6), and we will neglect it below.

Now we introduce spherical coordinates, (λ, φ, r), where λ is longitude, φ is latitude, and r is the radial distance from the center of the Earth. The unit vectors in the (λ, φ, r) coordinates are $\mathbf{e}_\lambda$, $\mathbf{e}_\varphi$, and $\mathbf{e}_r$, respectively. The directions in which the unit vectors actually point depend on where you are. By studying a globe (or just imagining the spherical Earth), you should be able to see that the direction of $\mathbf{e}_\lambda$ depends on longitude and that the directions of $\mathbf{e}_\varphi$ and $\mathbf{e}_r$ depend on both longitude and latitude. Of course, the *magnitudes* of the unit vectors are spatially constant.

Using these unit vectors, the velocity vector can be written as

$$\mathbf{V} \equiv u\mathbf{e}_\lambda + v\mathbf{e}_\varphi + w\mathbf{e}_r, \qquad \text{(A4.7)}$$

where the zonal, meridional, and radial velocity components of the velocity vector are defined by

$$u \equiv r\cos\varphi \frac{D\lambda}{Dt}, \ v \equiv r\frac{D\varphi}{Dt}, \ w \equiv \frac{Dr}{Dt}. \qquad \text{(A4.8)}$$

Simple geometrical reasoning leads to the following formulas:

$$\frac{D\mathbf{e}_\lambda}{Dt} = \frac{D\lambda}{Dt}\sin\varphi\mathbf{e}_\varphi - \cos\varphi\frac{D\lambda}{Dt}\mathbf{e}_r$$
$$= \left(\frac{u\tan\varphi}{a}\right)\mathbf{e}_\varphi - \frac{u}{a}\mathbf{e}_r, \qquad (A4.9)$$

$$\frac{D\mathbf{e}_\varphi}{Dt} = \frac{D\lambda}{Dt}\sin\varphi\mathbf{e}_\lambda - \frac{D\varphi}{Dt}\mathbf{e}_r$$
$$= \left(\frac{u\tan\varphi}{a}\right)\mathbf{e}_\lambda - \frac{u}{a}\mathbf{e}_r, \qquad (A4.10)$$

$$\frac{D\mathbf{e}_r}{Dt} = \cos\varphi\frac{D\lambda}{Dt}\mathbf{e}_\lambda + \frac{D\varphi}{Dt}\mathbf{e}_\varphi$$
$$= \frac{u}{a}\mathbf{e}_\lambda + \frac{v}{a}\mathbf{e}_\varphi. \qquad (A4.11)$$

Using (A4.7)–(A4.11), Equation (A4.6) can now be written, in component form, as

$$\frac{Du}{Dt} - \left(2\Omega + \frac{u}{r\cos\varphi}\right)(v\sin\varphi - w\cos\varphi) = -\frac{\alpha}{r\cos\varphi}\frac{\partial p}{\partial\lambda} - F_\lambda,$$
$$\frac{Dv}{Dt} + \left(2\Omega + \frac{u}{r\cos\varphi}\right)u\sin\varphi + \frac{vw}{r} = -\frac{\alpha}{r}\frac{\partial p}{\partial\varphi} - F_\varphi,$$
$$\frac{Dw}{Dt} - \left(2\Omega + \frac{u}{r\cos\varphi}\right)u\cos\varphi - \frac{v^2}{r} + g = -\alpha\frac{\partial p}{\partial r} - F_r. \quad (A4.12)$$

The equations given in Equation (4.2), in Chapter 4, which describe the geostrophic wind, can be obtained from the first two equations of (A4.12) by neglecting most of the terms. This can be justified away from the Equator and above the frictional influence of the Earth's surface. The hydrostatic equation can be obtained as an approximation to the third equation of (A4.12), again by neglecting most of the terms. The hydrostatic approximation is very accurate, and for many (but not all) purposes it can be used as a replacement for the full equation of vertical motion.

FORCING AND RESPONSE

"FEEDBACK" IS A TERM THAT IS WIDELY USED IN MANY technical fields, and even in daily life. Positive feedbacks amplify and negative feedbacks regulate or damp.

A familiar example of a positive feedback is the unpleasant squeal produced by a sound system that includes a microphone, an audio amplifier, and a speaker. A sound picked up by the microphone is amplified and played through the speaker. If the speaker is too close to the microphone or if the gain of the amplifier is turned up too high, the sound produced by the speaker is picked up by the microphone and further amplified, leading to the familiar screeching of audio feedback. Small perturbations are amplified; this is characteristic of a positive feedback.

A familiar example of a negative feedback is temperature control by a thermostat that is coupled to a heating, ventilation, and air conditioning (HVAC) system. A target temperature is preset by the user. The thermostat detects temperature excursions from the target value and commands the HVAC system to fight against those excursions by warming when the air is too cool and cooling when the air is too warm. Small perturbations are damped; this is characteristic of negative feedback.

Both positive and negative feedbacks occur in the climate system. Positive feedbacks tend to increase the system's variability, while negative feedbacks reduce it. It is important to distinguish between *internal* and *external* processes. For the case of climate feedback, internal processes are those that occur within the climate system itself. Examples include weather systems, ocean circulations, and land-surface processes. External processes are unaffected by the state of the system. As an example, the flow of energy from the Sun is unaffected by the state of the Earth's climate system. There is a gray area. For example, human activities are often considered external to the climate system, even though human affairs are strongly affected by climate, and even though humans are now capable of altering the climate. At some level, humanity is a part of the climate system.

The climate system is *forced* by external processes. For example, as explained in Chapter 2, the Earth's temperature is partly determined by the Sun's energy output. This is a forcing. Changes in forcing are due to changes in the external processes. A change in the brightness of the Sun could produce, as a *response*, a change in the Earth's temperature. In Chapter 2, we showed that

$$\left(\frac{\Delta T_S}{T_0}\right) \cong \left(\frac{\Delta S}{S_0}\right) - 16\left(\frac{\Delta\alpha}{1-\alpha_0}\right) - 4\frac{\Delta\varepsilon}{\varepsilon_0}. \tag{5.1}$$

Table 5.1
Classification of the changes that appear in Equation (5.1).

Quantity	Meaning	Classification
ΔS	Change in solar output	External
$\Delta \alpha$	Change in planetary albedo	Partly internal and partly external
ΔT_s	Change in surface temperature	Internal; this is the "response" of the system
$\Delta \varepsilon$	Change in bulk emissivity	Partly internal and partly external

Recall that ΔS, $\Delta \alpha$, ΔT_s, and $\Delta \varepsilon$ are the changes in the solar output, the planetary albedo, the surface temperature, and the bulk emissivity, respectively, as the climate system passes from one equilibrium to another, in response to a change in the external forcing. The various changes that appear in Equation (5.1) can be classified as external or internal, as shown in Table 5.1. A change in the solar output is clearly external. A change in the planetary albedo is usually assumed to be internal and can result from changes in snow and ice cover, vegetation, or cloudiness. The change in the albedo was assumed to be zero in Chapter 2. An externally forced change in the planetary albedo can occur in response to volcanic eruptions, if we agree to consider such eruptions as external to the climate system. It would also be possible to produce external changes in the planetary albedo by filling the stratosphere with reflective aerosols as a "geo-engineering" approach to counteract global warming.

The change in the surface temperature is internal, and in this discussion it is treated as the response of the system to a change in external forcing. A change in the bulk emissivity can arise from external causes, for example as a result of an anthropogenic increase in atmospheric CO_2 concentration. It can also have an internal component, however, due to changes in water vapor, clouds, and the lapse rate, as discussed below.

THE SNOW AND ICE ALBEDO FEEDBACK

The climate feedback due to changes in snow and ice cover is not really an atmospheric feedback, but it is particularly simple, so we'll use it as an introductory example.

In a warmer climate, snow and ice cover are expected to decrease (although there can be exceptions to this). The dependence of snow and ice cover on surface temperature is the first essential ingredient of the snow and ice albedo feedback. A second essential ingredient is that snow and ice absorb less solar radiation than the darker surfaces that they cover up. When dark soil or rock emerges from beneath white snow or a retreating glacier, or when the dark ocean emerges from beneath the white, sometimes snow-covered sea ice, more solar radiation is absorbed by the newly darkened surface. We can write

$$\Delta \alpha = \left(\frac{\partial \alpha}{\partial T_S} \right)_{\text{ice}} \Delta T_S, \tag{5.2}$$

where $\left(\frac{\partial \alpha}{\partial T_s}\right)_{\text{ice}}$ is the rate at which the albedo contribution associated with snow and ice changes, in response to a change in the surface temperature. The *partial* derivative indicates that the change in surface temperature is considered to occur while other parameters, such as cloudiness, remain fixed. We expect the albedo to decrease as the surface temperature increases, so that $\left(\frac{\partial \alpha}{\partial T_s}\right)_{\text{ice}} < 0$. The reduced albedo leads to an increase in the absorption of solar radiation. The added solar energy causes warming, which, in turn, can cause even more melting. This is a positive feedback "loop," as depicted in Figure 5.1.

The positive snow and ice albedo feedback can amplify a cooling, as well as a warming. If the climate cools, the resulting increased snow and ice cover reflects more sunlight back to space, and the cooling is amplified. Both positive and negative feedbacks can affect the response of the system to forcings of either sign.

To see how the snow and ice feedback affects the response of the surface temperature to a change in the external forcing, we substitute (5.2) into (5.1), and solve for ΔT_S. Here we are temporarily assuming that the albedo changes *only* as a result of melting snow and ice. The result is

$$\left(\frac{\Delta T_S}{T_0}\right) \cong \frac{\left(\dfrac{\Delta S}{S_0}\right) - 4\dfrac{\Delta \varepsilon}{\varepsilon_0}}{1 + \left(\dfrac{16 T_0}{1 - \alpha_0}\right)\left(\dfrac{\partial \alpha}{\partial T_S}\right)_{\text{ice}}}. \tag{5.3}$$

In the numerator on the right-hand side of (5.3), we have the changes in the external forcing due to the Sun and

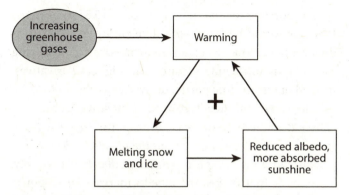

Figure 5.1. The positive feedback loop due to the dependence of snow and ice cover on temperature.

Here we show the feedback resulting from an external perturbation, that is, an increase in greenhouse gases. In this case, the positive feedback leads to increased warming. It is also possible, however, for the same physical mechanism to amplify a cooling, which might be externally forced, for example, by a decrease in the Sun's energy output.

the bulk emissivity. In the denominator, we have *a negative contribution* due to the negative value of the partial derivative $\left(\frac{\partial \alpha}{\partial T_s}\right)_{ice}$. Because this term makes the denominator on the right-hand side of (5.3) smaller, it increases the response, as measured by $\left(\frac{\Delta T_s}{T_0}\right)$.

Notice that we have made an approximation in deriving (5.3). The quantity $\left(\frac{\Delta \alpha}{\Delta T_s}\right)_{ice}$ has been replaced by $\left(\frac{\partial \alpha}{\partial T_s}\right)_{ice}$. This approximation is useful when the changes in the albedo and the surface temperature are sufficiently small. With larger changes, higher order derivatives would be needed for accuracy.

Is it possible for the denominator of (5.3) to be zero? If so, we would be faced with division by zero, which is unacceptable. The issue can be avoided by generalizing the analysis to include second (and higher) derivatives, instead of just the first partial derivative as shown in (5.3).

The snow and ice feedback is positive because it causes the albedo to decrease in response to a warming. In fact, *any feedback that causes the albedo to decrease in response to a warming is a positive feedback.* Conversely, a feedback that makes the albedo increase in response to warming is a negative feedback. As discussed below, cloud feedbacks can make the albedo change and can be either positive or negative.

Similarly, *any feedback that makes the bulk emissivity decrease in response to a warming is a positive feedback.* The positive water vapor feedback, discussed below, is the most important example. Cloud feedbacks can also cause the bulk emissivity to change in response to a warming and can be either positive or negative.

The approach to feedback analysis outlined above was developed for use in engineering (e.g., Bode, 1975) and was first applied to the problem of climate change by Schlesinger (1989).

THE WATER VAPOR FEEDBACK

The positive water vapor feedback is the most important atmospheric feedback in the climate system. Its existence was recognized over a century ago (Arrhenius, 1896). In the early days of computer-based climate modeling,

Manabe and Wetherald (1967) developed the concept in a modern context.

The basic idea is very simple: Recall from Figure 3.4 that the saturation vapor pressure of water increases exponentially with temperature. At the Earth's globally averaged surface temperature of 288 K, the saturation vapor pressure increases by about 7% for a 1 K warming. There is a tendency for the relative humidity of the air to remain approximately constant as the climate changes. It follows that the water vapor content of the atmosphere will increase by about 7% per K as the climate warms. Because water vapor is a powerful greenhouse gas, that is, it strongly absorbs and emits infrared radiation, an increase in the atmospheric water vapor content causes an increase in the downwelling infrared radiation at the Earth's surface. This favors a further warming of the oceans. The initial perturbation is amplified.

Surface evaporation can moisten a deep layer of air only if a mechanism exists to carry water vapor upward away from the surface. Boundary-layer turbulence helps, but only through the depth of the boundary layer, which is typically less than 1 km. Further lifting of moisture, beyond the boundary-layer top, is necessary if the total moisture content of the atmospheric column is to increase by a lot. The most important mechanism for such further lifting is cumulus convection. For this reason, convective clouds play an essential role in the water vapor feedback.

The mass of water vapor in the atmosphere exceeds the mass of carbon dioxide by about a factor of 4, and water vapor contributes more to the downward infrared at the Earth's surface than CO_2 does. Despite these facts, CO_2 plays a controlling role in the greenhouse effect, while water vapor plays a subservient role. The reason is that water vapor is condensible and can be removed from the atmosphere by precipitation, whereas CO_2 does not condense under conditions found in the Earth's atmosphere.

Lacis et al. (2010) illustrated the primary role of CO_2 through a clever and simple experiment with a climate model. When run with realistic present-day CO_2 concentrations, the model produces a realistic simulation of today's observed climate. In the experiment, Lacis and colleagues removed all of the CO_2 (and other noncondensing greenhouse gases) from the atmosphere. The results were spectacular. The surface temperature began to fall immediately as a direct result of the absence of CO_2. This led to a reduction in the water vapor content of the atmosphere and additional cooling through the water vapor feedback. The positive snow and ice albedo feedback also reinforced the cooling. The dramatic results are shown in Figure 5.2. Within 10 simulated years, the Earth approached an ice-covered state in which the frigid model atmosphere contained very little water vapor.

Suppose that we ran a "mirror-image" experiment, in which the CO_2 concentration was realistic but the water vapor content of the atmosphere was set to zero at the beginning of the simulation. Evaporation from the oceans would restore a realistic water vapor distribution within a few simulated weeks, and the simulated climate would stay close to what we observe today.

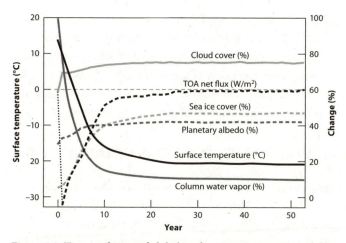

Figure 5.2. Time evolution of global surface temperature, top-of-the-atmosphere net radiation, column water vapor, planetary albedo, sea ice cover, and cloud cover, after the zeroing out of the CO_2 and other noncondensing greenhouse gases.

The initial conditions are for a preindustrial atmosphere. Surface temperature and top-of-the-atmosphere net radiation use the left-hand scale.

Source: Lacis et al. (2010).

Although we humans do add water to the atmosphere, for example as a side effect of agricultural irrigation, there is no danger that our civilization will substantially increase the water vapor content of the atmosphere because condensation and precipitation limit the water vapor increase.

The water vapor feedback acts on time scales of weeks, so we can see its effects without waiting for climate change. In a very interesting experiment, Hall and

Manabe (1999) showed that the internal (unforced) variability of a climate model becomes unrealistically weak when the water vapor feedback is artificially suppressed.

An extreme version of the water vapor feedback, called the "runaway greenhouse," is believed to have played a key role in the evolution of the atmosphere of Venus.

HOW MULTIPLE FEEDBACKS COMBINE

Now we have two feedbacks, one due to ice and snow, and the other due to water vapor. This gives us an opportunity to ask a simple but important question: How do feedbacks combine? The answer is that they add, but in a not-so-obvious way.

The bulk emissivity of the atmosphere, ε, is due to the combined effects of CO_2, water vapor, other greenhouse gases, and clouds. In Chapter 2, we considered a perturbation to ε due to doubling CO_2, and with the assumption of no feedbacks. In order to modify our analysis to take into account the water vapor feedback, we have to divide the bulk emissivity into a part due to CO_2 and a part due to water vapor:

$$\varepsilon = \varepsilon_{CO2} + \varepsilon_{H2O}. \tag{5.4}$$

After some algebra, we find, corresponding to Equation (5.3), that

$$\left(\frac{\Delta T_S}{T_0}\right) \cong \frac{\left(\frac{\Delta S}{S_0}\right) - 4\frac{(\Delta \varepsilon)_{CO2}}{\varepsilon_0}}{1 + \left(\frac{16 T_0}{1 - \alpha_0}\right)\left(\frac{\partial \alpha}{\partial T_S}\right)_{ice} + \left(\frac{4T}{\varepsilon_0}\right)\frac{\partial \varepsilon_{H2O}}{\partial T_S}}. \tag{5.5}$$

The forcing due to a change in CO_2 concentration appears in the numerator of Equation (5.5), through the quantity $(\Delta\varepsilon)_{CO2}$. The water vapor feedback appears in the denominator, through the partial derivative $\frac{\partial\varepsilon_{H2O}}{\partial T_s}$.

Notice that the term representing the water vapor feedback *is simply added* to the term representing the snow and ice albedo feedback. We will not introduce any more mathematical complications here, but suffice it to say that including even more feedbacks, for example due to clouds or changes in the lapse rate, would simply add more partial-derivative terms in the denominator of (5.5). This is the sense in which feedbacks add. Including more feedbacks makes the equation more complicated because there are more terms, but the basic form of the equation does not change. That's a nice, simple result.

Alright then, let's add some more feedbacks.

CLOUD FEEDBACKS

Cloud feedbacks are infamous. Their potential importance has been recognized at least since the 1970s (Schneider, 1972; Arakawa, 1975; Charney, 1979). For decades, they have been highlighted as a major source of uncertainty in predictions of future climate change. Why is the problem taking so long to solve?

Although cloud processes are among the most important for climate, they are also among the most difficult to understand and predict. As discussed in Chapter 3, clouds usually form in rising air, which expands and cools until condensation occurs. Cloud processes can be grouped into five categories, all of which are important for the climate system:

Cloud microphysical processes, involving cloud drops, ice crystals, and aerosols, with scales on the order of microns

Cloud dynamical processes associated with the motion of the air in cumulus clouds, with scales on the order of a few kilometers

Cloud radiative processes, involving the flows of both solar and terrestrial radiation through the cloud field

Cloud turbulence processes, which include boundary-layer (near-surface) turbulence, the entrainment of environmental air into the clouds, and the formation of small clouds

Cloud chemical processes, which, among other things, are important for the annual formation of the ozone hole through the destruction of ozone in polar stratospheric clouds

These various processes interact with each other on space scales of a few kilometers and time scales of a few minutes, and they also jointly interact with larger-scale weather systems, up to the size of the Earth itself.

Clouds also couple climate processes together. For example, they cast shadows on the land and ocean, which tend to reduce the surface temperature and evaporation rate.

Each cloud process has the potential to feed back as the climate state evolves due to externally forced variability. As a result, there are many different kinds of cloud feedbacks. Broadly speaking, they occur through changes in cloud amount, cloud top height, and cloud optical properties. In a climate change, there can be many different changes in the geographical patterns and seasonal distributions of both high and low clouds. The net cloud feedback results from the combined effect of these various changes.

Two examples of cloud feedbacks are discussed below. Many others are discussed in the literature.

Low-Cloud Feedback

Optically thick clouds behave like blackbodies. Low clouds emit at relatively warm temperatures, not very different from the temperature of the Earth's surface. As a result, an increase in low cloud amount has relatively little effect on the OLR.

On the other hand, low clouds can be very bright (as seen from above), reflecting back to space up to half of the solar radiation that hits them. As a result, they tend to cool the Earth, in the present climate. In a future climate warmed by increasing greenhouse gases, an increase in low cloud amount would increase the cooling, and so could reduce the warming. On the other hand, a decrease in low-cloud amount would increase the warming.

Low cloud amount tends to be greater when the sea-surface temperature (SST) is cooler than the average SST at that latitude. For example, the SST west of San Diego is cooler than the average SST at 32°N, and low clouds are

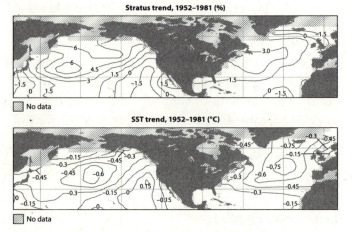

Figure 5.3. Observed trends of stratus cloud amount (top) and sea surface temperature (bottom), for the period 1952–81.

Source: The figure was provided by Joel Norris. It is based on work published by Norris and Leovy (1994).

very abundant there, especially in the northern summer. Over cold water, low clouds typically take the form of a uniform overcast. When the water is warmer, the low clouds change their type, to shallow cumulus clouds with smaller cloud fractions.

Figure 5.3 shows a possible example of an observed (Norris and Leovy, 1994) cloud feedback. The two panels of the figure show observed trends in SST and stratocumulus cloudiness, over a 30-year period. Downward trends in sea surface temperature are colocated with upward trends in stratocumulus amount, and vice versa. There are at least two possible interpretations of these correlated trends in SST and stratocumulus cloud amount, which do not contradict each other. The first is that a cooling (warming) of the sea favors an increase (decrease) in low cloud amount; this is plausible in light of our understanding of the physics of low-level marine clouds. The second interpretation is that an increase (decrease) in stratus cloud amount favors a decrease (increase) in the sea surface temperature because the clouds reflect solar radiation that would otherwise be absorbed by the ocean.

The positive cloud feedback suggested in Figure 5.3 is called a shortwave cloud feedback because it primarily involves solar radiation. Clement et al. (2009) provide further observational evidence for a positive shortwave cloud feedback associated with low clouds.

The discussion of feedbacks in this chapter is formulated mostly in terms of globally averaged quantities. In practice, however, we are interested in analyzing the external forcing, the response, and the feedbacks *as spatially distributed fields*, like those shown in Figure 5.3.

High-Cloud Feedback

High, cold cirrus clouds are often somewhat transparent, so they don't reflect as much solar radiation as typical low-level clouds, but they do efficiently absorb the infrared radiation coming from the Earth's surface, leading to a radiative warming near the cloud-base level. The high clouds emit to space, but only weakly because of their very cold temperatures. They therefore tend to warm the Earth as a whole. An increase in high cloud amount will tend to enhance greenhouse warming, while a decrease will tend to reduce it.

Even if the high-cloud amount remains the same, the high clouds can produce a positive feedback as the surface warms up, because they will absorb the increased infrared coming from the warmer surface. Tropical cumulus clouds often stop about 15 km above the Earth's surface, where the temperature is about 200 K. Hartmann and Larson (2002) suggested that the reason is that at a temperature of 200 K the infrared radiative cooling of the atmosphere becomes weak. The cooling becomes weak because the water vapor concentration is very small. The air is dry because it is so cold that the saturation specific humidity of water vapor is very small.

Radiative cooling aloft tries to steepen the lapse rate and generates gravitational potential energy. Warming produced by the cumulus clouds balances the radiative cooling. When the radiative cooling stops, the cumulus cloud warming must also become weak. According to Hartmann and Larson, this is why the cumulus clouds top out at the level where the temperature is 200 K.

Taking this idea a step further, Hartmann and Larson hypothesized that if the climate changes in such a way that the height where the temperature is 200 K moves up or down, the tops of the cumulus clouds will move up and down too, so that the cloud-top temperature is always 200 K. This is called the "fixed anvil temperature" (FAT) hypothesis.

The tall tropical cumuli produce broad regions of optically thick cirrus and anvil clouds near their tops, which therefore emit infrared at a temperature of about 200 K. The FAT hypothesis of Hartmann and Larson implies that this will be true even in a different climate. It follows that in regions of deep cumulus convection, the OLR will not change even when the climate changes. The OLR will therefore be insensitive to the surface temperature. The Earth will not be able to increase its infrared emission to space as the surface temperature increases. The bulk emissivity will therefore decrease as the Earth warms. This is a positive cloud feedback. It is called a longwave cloud feedback because it primarily involves infrared radiation. Further discussion is given by Zelinka and Hartmann (2010).

THE LAPSE-RATE FEEDBACK

A climate change in which the warming increases with height is one in which the lapse rate decreases. The change in the lapse rate is a feedback.

An infrared photon that is emitted in the upper atmosphere can escape to space more easily than one that is emitted near the surface. Suppose that the atmosphere warms up, so that the overall rate of emission increases. If the warming is mostly in the upper troposphere, the additional thermal energy be radiated away to space easily, and *the bulk emissivity increases*. If the warming is mostly in the lower troposphere, the energy has a harder time getting out to space, and the bulk emissivity decreases.

For a very simple reason, explained below, there is a tendency for the lapse rate to become weaker as the surface temperature warms. This is especially true in the tropics. If the surface temperature warms, the lapse rate decreases, and the upper tropospheric temperature warms even more. The bulk emissivity increases, and the warming is damped. This is a *negative* longwave feedback.

The reason why the lapse rate decreases as the temperature warms can be seen in Figure 3.5, which shows the variations of the moist adiabatic lapse rate with temperature and pressure. For a given pressure, the moist adiabatic lapse rate decreases as the temperature increases. This follows directly from the thermodynamic properties of water. As discussed in Chapter 3, the actual lapse rate of the tropical troposphere, and also to some extent the summer midlatitude troposphere, approximates the moist adiabatic lapse rate. If the surface temperature warms, the lapse rate decreases, and the upper troposphere warms even more.

The lapse rate feedback is negative because it tends to damp changes in the surface temperature. It is important to realize that the feedbacks themselves are manifestations of climate change. Even a negative feedback, like a decrease in the lapse rate, is a climate change.

OBSERVING FEEDBACKS IN THE CLIMATE SYSTEM

There are many additional feedbacks in the climate system. They involve the large-scale atmospheric circulation, the ocean circulation, vegetation, ice sheets, and more. Some feedbacks, such as melting of the continental ice sheets, are expected to occur on time scales of centuries, while others, such as the water vapor feedback, occur on time scales of days or weeks.

Feedbacks can be measured through their influences on the variability of a system, that is, by watching the system fluctuate. Feedbacks on climate change can be observed by measuring the changes that occur over extended periods of time (decades or longer). Of course, we would like to measure the various feedbacks right now. The fast feedbacks can affect externally forced changes on short time scales that are directly observable now. These include volcanic eruptions, El Niño events, the seasonal cycle, storms, and the diurnal cycle. By observing the role of feedbacks in the response of the climate system to such rapid changes in external forcing, we can test our ideas about fast feedbacks. This will help us to understand how feedbacks work in the more ponderous process of climate change over decades and centuries.

Fifty or a hundred years from now, we will be able to compare observations of ice and snow, water vapor, cloudiness, and other internal variables of the climate system with corresponding data from today. The differences will represent the net feedback on the climate change that occurs between now and then. It may be difficult to separate the net feedback into distinct contributions from the various internal processes that we humans like to talk about, but that is not really a problem. It is only the *net* feedback that matters, and the net feedback will be readily observable. Our descendants will be able to tell whether or not we got it right.

THE WATER CYCLE

As discussed in earlier chapters, the energy cycle of the climate system begins with the absorption of solar radiation, continues as energy is transformed and transported inside the atmosphere and oceans, and ends as terrestrial radiation is emitted back to space.

We can also follow the flow of water through the system. The evaporation of water from the surface must be balanced, somewhere on Earth, by precipitation falling back to the surface. This is the water cycle, or hydrologic cycle. It is intimately coupled to the energy cycle. It is no exaggeration to say that the Earth's climate is strongly controlled by the interactions of water and energy.

Earlier chapters have discussed the effects of water vapor and clouds and the flows of solar and terrestrial radiation. We have explored the importance of latent heat as a component of the total energy of the air. We have discussed the key role of cumulus convection, which is fueled by latent heat release, in transporting energy upward. We have described the strongly positive water vapor feedback that amplifies climate changes.

Meanwhile, the ocean also plays a significant role in poleward energy transport, especially in the tropics and

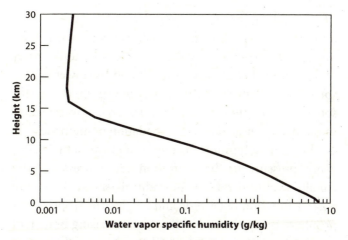

Figure 6.1. The vertical profile of the globally averaged specific humidity.

Note that the horizontal scale is logarithmic. In the lower tropo-sphere, tropical values are larger than plotted here, and high-latitude values are smaller.

subtropics. In addition, the ocean serves as an enor-mous reservoir of thermal energy. The Earth's albedo is strongly influenced by sea ice, and snow and ice on the land surface.

Finally, we should not forget the great importance of water for life, and the many ways that biology affects the composition of the atmosphere and other aspects of the climate system.

In short, water is at the center of the Earth's climate system. In this chapter, we explore some of the ways that water influences climate, including additional feedbacks on climate change.

The globally averaged vertical profile of the specific humidity (water vapor mass fraction) is shown in Figure 6.1. The specific humidity decreases by four orders of magnitude between the surface and the lower stratosphere. It actually increases weakly upward in the stratosphere because there is a chemical source of water vapor in the stratosphere due to the oxidation of methane.

The relative humidity is the ratio of the actual water vapor pressure to the saturation vapor pressure. The zonally averaged relative humidity is shown in Figure 6.2 as a function of latitude and height. The tropical troposphere is very humid in the (shared) rising branch of the Hadley cells (i.e., the ITCZ), and high relative humidities are also found in middle latitudes, especially in winter, where strong low-pressure systems are frequent. A maximum appears in the tropical upper troposphere, near the level where deep cumulus clouds reach their level of neutral buoyancy and discharge their humid contents. The maximum shifts northward in July, and southward in January, following the movement of the rising branch of the Hadley cells. The subtropical minimums are associated with the sinking branches of the Hadley cells.

Figure 6.3 shows maps of the precipitation rate for January and July. The right-hand panels show the corresponding zonally averaged surface precipitation rates. Precipitation tends to be very "spotty" in both space and time. As a result, average values are difficult to determine accurately. The zonally averaged precipitation has its strongest maximum in the tropics, with secondary

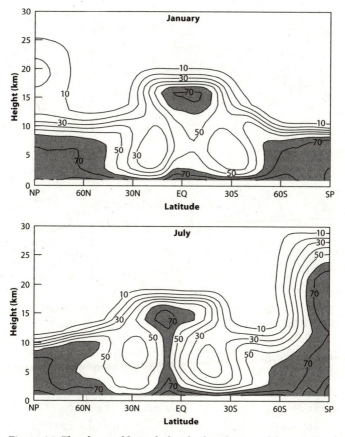

Figure 6.2. The observed latitude-height distribution of the zonally averaged relative humidity (in %)

Values larger than 60% are shaded.

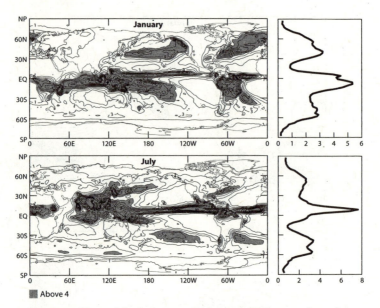

Figure 6.3. Maps of the January and July precipitation rate, from the Global Precipitation Climatology Center (Adler et al., 2003).

The contour interval is 1 mm day^{-1}. Values greater than 4 mm day^{-1} are shaded. A layer of water 1 mm deep, covering 1 m^2, has a mass of 1 kg, so 4 mm day^{-1} is equivalent to 4 kg m^{-2} day^{-1}.

maximums in the middle latitudes. The tropical maximum is associated with the Intertropical Convergence Zone and the monsoons, while the middle latitude maximums are associated with baroclinic wave activity in the winter and monsoon circulations in the summer. There are minimums in the subtropics, where the major deserts occur. The global mean of the precipitation rate is a little less than 3 mm day^{-1}.

Figure 6.3 makes it clear that the rainiest regions of the world are in the tropics. The ITCZ, discussed in Chapters 3 and 4, is clearly visible as an east-west "stripe" over the tropical Pacific and Atlantic Oceans, most prominent in July. There are major seasonal shifts in the locations of the tropical precipitation. The precipitation follows the sun, shifting north in July and south in January. In January, the strongest zonally averaged precipitation is south of the Equator; it appears not as a continuous east-west stripe but as a series of maximums at various longitudes. For example, heavy rain falls over the Amazon basin, southern Africa, the Indian Ocean, the maritime continent north of Australia, in the South Pacific Convergence Zone that extends southeastward from the intersection of the Date Line with the Equator, and also across most of the tropical Pacific and Atlantic Oceans north of the Equator. In July, heavy rainfall occurs in the extreme northern part of South America, in the neighboring Caribbean Sea and tropical North Atlantic Ocean, over India and neighboring regions of southeast Asia, and to the north of the maritime continent, off the east coast of tropical Asia.

Precipitation minimums are conspicuous in the subtropics, especially on the winter side of the Equator, where the subsiding branch of the main Hadley cell brings dry air down from the upper troposphere. The precipitation minimums are particularly strong in the vicinity of the subtropical high-pressure cells, which can be seen in the sea level pressure maps presented in Chapter 4.

In middle latitudes, especially over the Northern Hemisphere oceans, precipitation tends to be stronger in winter. The rain and snow of the midlatitude winter are produced by the winter storms discussed in Chapter 4. There is a tendency for such storms to form near the east coasts of North America and Asia, where horizontal temperature contrasts are particularly strong. The storms then move eastward and poleward over the oceans, in what are called "storm tracks," which coincide with the precipitation maximums of the North Pacific and North Atlantic, as seen in the upper panel of Figure 6.3. Recall from the sea level pressure map of Chapter 4 that there are climatological low-pressure centers near Alaska and Iceland in winter.

As is clear from Figure 6.3, minimums of the zonally averaged precipitation occur in the subtropics and secondary maximums occur in the middle latitudes. Mid-latitude precipitation is also highly variable. For example, the precipitation over northern Asia occurs mainly in the summer. The warm currents off the east coasts of North America and Asia receive heavy precipitation mainly in winter. The northwestern portion of the United States receives heavy precipitation in winter but not in summer. The figure also shows the regions that receive plentiful precipitation throughout the year include eastern North America, the extreme southern tip of South America, and England.

The zonal means of the annually averaged evaporation, precipitation, and their difference are shown in Figure 6.4. The evaporation rate has maximums in the

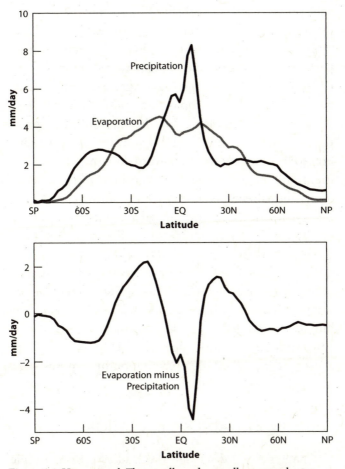

Figure 6.4. Upper panel: The zonally and annually averaged rates of evaporation and precipitation. Lower panel: The zonally and annually averaged difference between the rates of evaporation and precipitation. Positive values mean that evaporation is larger than precipitation.

subtropics, where the reliable trade winds blow across huge expanses of moderately warm ocean water.

The evaporated water is carried by the winds to other parts of the world, where precipitation rains it out. The figure shows that the zonally averaged precipitation rate has maximums in the tropics and middle latitudes, both equatorward and poleward of the subtropical evaporation maximums. This implies that water vapor is transported from the subtropics toward both the Equator and the middle latitudes, and this can be confirmed using direct measurements of the winds and the specific humidity. The flow of water toward the Equator is carried out by the lower branch of the Hadley cell, which was discussed in Chapter 4. The flow of water toward middle latitudes is due to baroclinic waves, in which warm, humid air moves poleward and upward, while colder, drier air moves equatorward and downward. When the humid air is lifted, it cools by expansion, promoting condensation and, ultimately, precipitation. This is the cause of the midlatitude precipitation maximums, seen in Figure 6.3.

THE HADLEY-WALKER CIRCULATION

In Chapter 4, we discussed the Hadley cells as simple overturning circulations in the latitude-height plane. The reality is more complicated. We did briefly mention, in Chapter 4, that the Hadley cells are particularly strong at some longitudes, for example, near the Asian monsoon. An additional complication is that the overturning can have an east-west component as well as the north-south "Hadley" component.

The most famous of the east-west overturning circulations is the "Walker Cell," named for the British physicist Gilbert Walker, who lived from 1868 to 1958. The Walker Cell, which is elegantly described by Philander (1990), has its rising branch over the warm waters of the western Pacific Ocean near New Guinea and its sinking branch over the cold water west of South America, near the Galapagos Islands. The Walker Cell thus spans the very considerable width of the Pacific Ocean.

As you can probably imagine, the east-west overturning and the north-south overturning cannot really be cleanly separated, so we often speak of the "Hadley-Walker circulation," combining the two overturnings into a single component of the tropical atmospheric circulation. A cartoon of the Hadley-Walker circulation is shown in Figure 6.5. There is a lot going on in this figure. The horizontal axis could, for example, represent distance along a line that stretches from southwest near New Guinea (on the left side of the sketch) to northeast near Southern California (on the right side of the sketch). The sea surface temperature (SST) is warm (~302 K) near New Guinea and cold (~287 K) near California, for reasons that will be discussed in Chapter 8. As a matter of convention, the west (and equatorward) side of the region is commonly called the *Warm Pool*, and the east (and poleward) side can be called the *Cold Pool*.

In the convectively active Warm Pool, the warm and humid near-surface air rises through cumulus towers

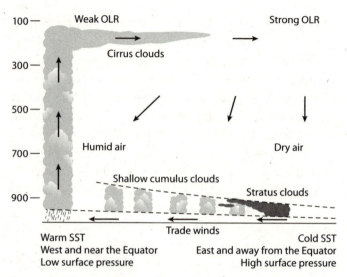

Figure 6.5. Schematic diagram showing the relationship of horizontal and vertical motions, cloud types, OLR, and SST, in the Hadley-Walker circulation of the Northern Hemisphere tropics and subtropics.

Source: Adapted from Schubert et al. (1995) and Pierrehumbert (1995).

made possible by latent heat release. The deep cumulus clouds warm and moisten the air column and produce abundant upper-tropospheric cloudiness. Above the Warm Pool, the OLR is relatively small because both the water vapor and the upper-tropospheric cloudiness absorb the infrared upwelling from near the warm surface and reemit at colder temperatures. Solar radiation is also reflected back to space by the deep convective cloud systems of this tropical region. The Warm Pool troposphere experiences a net radiative warming, primarily due to the blocking of upwelling longwave radiation from the surface and the lower troposphere.

Cold upper tropospheric air flows outward from the Warm Pool toward the Cold Pool. The air is dry because it is cold. It gradually subsides over the central and eastern Pacific, bringing very dry upper tropospheric air down into the lower and middle troposphere of the Cold Pool region. The sinking air then flows back toward the Warm Pool in the low-level trade winds, gradually warming and moistening as it travels westward and equatorward. Near the Equator, where the effects of the Earth's rotation are weak, the air tends to flow from high pressure to low pressure. The warm SST of the Warm Pool favors low surface pressure, and the cold water of the Cold Pool favors higher surface pressure. The pressure force thus pushes the near-surface air from the Cold Pool to the Warm Pool, and this is consistent with the trade-wind flow shown in the diagram.

Shallow, low-level clouds are found over the central and eastern Pacific, where the lower troposphere is dry, cool, and subsiding. In the Cold Pool, low-level "stratus" clouds form because the water vapor evaporated from the ocean is trapped in a thin layer near the surface. The stratus clouds cover the sky uniformly, and reflect back to space about half the sunlight that hits them, so they strongly reduce the solar warming of the ocean below. This helps to maintain the cooler SSTs of the Cold Pool. Stratus clouds form preferentially over cold water, so a synergy is at work here.

Over the central Pacific, where the ocean is warmer and the atmospheric subsidence is weaker, shallow cumulus clouds predominate. They cover only about 20% of the sky and produce a modest amount of precipitation. They carry moisture upward, enabling surface evaporation to moisten a layer of air 2 to 3 km deep. This moist layer flows into the Warm Pool, fueling the heavy precipitation there.

The shallow stratus and cumulus clouds and the dry, subsiding air above them emit upward at temperatures only slightly cooler than the surface, so the OLR over the Cold Pool is relatively strong. There is an interesting and counterintuitive point here, which was emphasized by Pierrehumbert (1995): *The Warm Pool emits weakly to space, while the Cold Pool emits strongly.*

HEATING AND VERTICAL MOTION

As discussed in Chapter 4, horizontal temperature gradients tend to be small in the tropics, where the weak Coriolis effect allows the winds to smooth out temperature and pressure differences as quickly as they appear. This is very different from middle latitudes, where the stronger Coriolis effect allows much stronger horizontal temperature contrasts, especially in winter.

Because the temperature is horizontally uniform throughout the tropics, local heating by radiation or condensation cannot be balanced by importing cooler air from the sides. The only way to balance the heating is for the heated air to rise toward lower pressure, so that expansion can reduce its temperature. Similarly, tropical atmospheric cooling (in the sense of removing energy, for example by emitting infrared radiation) must be balanced by sinking toward higher pressure. Throughout the tropics, heating and rising motions go together and cooling and sinking motions go together. If we assume a steady state and horizontal homogeneity, the change of the dry static energy with time, following a parcel, can be written, using the chain rule, as

$$\frac{Ds}{Dt} \cong \frac{\partial s}{\partial z}\frac{Dz}{Dt}$$
$$= \frac{\partial s}{\partial z}w \quad , \tag{6.1}$$

where $w \equiv \frac{Dz}{Dt}$ is the time rate of change of height following the parcel, that is, it is the vertical velocity of the parcel. The dry static energy equation can then be written as

$$\left(\frac{\partial s}{\partial z}\right)\rho w = Q, \tag{6.2}$$

where Q is the heating rate per unit volume. Recall from Chapter 3 that the dry static energy increases upward, so that $\frac{\partial s}{\partial z} > 0$. Equation (6.2) says that an upward mass flux ($\rho w > 0$) balances heating, and a downward mass flux ($\rho w < 0$) balances cooling.

Rising motion occurs in the Warm Pool, where cumulus convection and radiation combine to make $Q > 0$. The small-scale effect of the heating is to allow the dry static energy to increase upward, following a rising parcel. Sinking motion occurs in the Cold Pool, which is radiatively cooled at the rate of about 1.5 K per day. The effect of the radiative cooling is to allow the dry static energy to decrease downward, following a sinking parcel. For purposes of this discussion, the radiative cooling rate can be considered as "given," or externally imposed, essentially determined by the temperature sounding and the CO_2 and water vapor contents of the air. The temperature sounding and the water vapor content are determined by the cloud processes of the Warm Pool region, and the CO_2 concentration is well mixed throughout the atmosphere, as discussed in Chapter 1.

For the reasons discussed in Chapter 3, the abundant thunderstorms of the Warm Pool atmosphere ensure that the temperature decreases upward there following the moist adiabatic lapse rate. This determines $\frac{\partial s}{\partial z}$ in the Warm Pool, and we can call this a convectively imposed dry static energy sounding. But the horizontal uniformity of tropical temperatures implies that *this same convectively imposed dry static energy sounding is found in the Cold Pool.*

Now think about Equation (6.2), as applied to the Cold Pool. We have argued, just above, that $\frac{\partial s}{\partial z}$ in the Cold Pool is essentially determined by the convection far away in the Warm Pool, and that the radiative cooling rate in the Cold Pool is also externally imposed. The only other variable in Equation (6.2) is the vertical velocity, w. We are thus led to the conclusion that *the rate at which the air sinks in the Cold Pool is determined by the radiative cooling rate and the convectively imposed dry static energy sounding.*

The same conclusion applies fairly well even in the middle latitudes in summer, simply because horizontal temperature gradients are weak there as well. It definitely does not apply in the midlatitude winter, where horizontal advection, in the presence of intense horizontal temperature contrasts, produces strong temperature changes.

The total mass of air subsiding over the Cold Pool is essentially the density times the vertical velocity times the width of the Cold Pool. This downward mass flow into the lower troposphere of the Cold Pool must be balanced by an outflow from the Cold Pool through the trade winds, which feeds an upward mass flux in the Warm Pool, which in turn supplies the subsiding air over the Cold Pool. Mass is conserved. We say that the mass budget must balance.

Similarly, the total rate of radiative energy loss over the Cold Pool is the radiative cooling rate times the width of the Cold Pool. To first order, the total radiative energy loss from the Cold Pool must be balanced by latent heat release (and possible radiative warming) over the Warm Pool. The relative widths of the Cold Pool and Warm Pool must adjust so that this overall energy balance is maintained.

The water budget must also balance, of course. The precipitation in the Warm Pool is made possible by the evaporation in the Cold Pool, which depends on the strength of the trade winds, and the SST and width of the Cold Pool. Again, the widths of the Warm and Cold Pools must adjust to achieve a balance.

Finally, the discussion above has been simplified by omitting any consideration of the angular momentum budget, which plays a very important role in the meridional (Hadley) component of the circulation. In a more complete analysis, the angular momentum budget must also balance.

WHAT DETERMINES THE SPEED OF THE WATER CYCLE?

The ideas of energy and water balance can be extended to the whole Earth and provide some insight into what determines the globally averaged rates of surface evaporation and precipitation, which are measures of the "speed" of atmospheric branch of the water cycle, that is, the rates at which water enters and leaves the atmosphere. Consider together the following points, which were made in earlier chapters:

For the Earth's surface, the globally averaged net radiative heating approximately balances the globally averaged evaporative cooling.

For the atmosphere, the globally averaged net radiative cooling approximately balances the globally averaged latent-heat release. The latent heat is supplied by the evaporation of water from the surface.

The clouds that affect the flows of solar and terrestrial radiation are themselves products of the hydrologic cycle. The clouds strongly modulate both the net surface radiative heating and the *net atmospheric radiative cooling*, which we will call the ARC. The latter can be defined as the net radiative loss of energy by the atmosphere due to solar and terrestrial radiation flowing across the atmosphere's top and bottom boundaries. In particular, the ARC is strongly affected by the high, cold cirrus clouds, many of which are formed within precipitating cloud systems. The cirrus clouds absorb the infrared radiation emitted by the warm atmosphere and surface below. The cirrus themselves emit much more weakly because they are very cold. This means that the cirrus effectively trap infrared radiation inside the atmosphere. For this reason, as the cirrus cloud amount increases, the radiative cooling of the atmosphere decreases.

Precipitating weather systems produce lots of cirrus clouds.

Taken together, these points suggest a negative feedback loop that regulates the strength of the hydrologic cycle.

To see how the negative feedback works, consider an equilibrium in which the ARC and latent heat release are in balance. Suppose that we perturb the equilibrium by increasing the speed of the hydrologic cycle, including the rate at which the atmosphere receives latent heat through evaporation from the oceans. The same perturbation increases the amount of high cloudiness, which reduces the rate at which the atmosphere is radiatively cooled. The atmosphere is now out of energy balance because the source of latent energy at the surface has increased and the radiative cooling rate has decreased.

In order to restore atmospheric energy balance, the hydrologic cycle has to slow down, that is, the initial perturbation has to be damped out. The required damping will occur quite naturally. A slowing of the hydrologic cycle is favored by the increased evaporative cooling of the surface, which will tend to reduce the surface temperature and, in turn, the evaporation rate. In addition, the reduced radiative cooling of the atmosphere, together with stronger cumulus convection, will tend to warm the upper troposphere, favoring a reduction in the intensity of cumulus convection and fewer high clouds. This is how the speed of the hydrologic cycle returns to normal and balance is reestablished.

The negative feedback loop described above tends to damp fluctuations of the hydrologic cycle, and because the energy and water cycles are linked, it also plays a role in regulating the energy cycle.

Although, for the reasons given above, the *globally averaged* fluctuations of the ARC and latent heat release are expected to be positively correlated over time, the geographical distributions of the ARC and latent heat release are negatively correlated. The explanation for this paradox is that the high clouds reduce the ARC near the places where precipitation is occurring.

THE WATER CYCLE OF A WARMER EARTH

Some current research aims at understanding how the hydrologic cycle will change on a warmer Earth. Human society is more vulnerable to changes in precipitation than to changes in temperature.

As mentioned in Chapter 3, at the Earth's globally averaged surface temperature of 288 K, the saturation vapor pressure increases by about 7% for a 1 K warming. There is a tendency for the relative humidity of the air to remain approximately constant as the climate changes. It follows that the water vapor content of the air near the surface will increase by about 7% per K, as the climate warms, and the upward slope of $e^*(T)$ (see Figure 3.3) becomes even steeper as the warming continues.

It is not obvious that a moistening of the air near the surface will lead to a moistening at higher levels. Cumulus convection carries moisture upward, however, and so it tends to create regions of high relative humidity in the air aloft. For this reason, a warming of the surface does in fact promote a moistening throughout the troposphere. A temperature increase of 3 K, expected by the end of the 21st century, can lead to about 25% more water vapor in the atmosphere. That's a huge change, especially considering that a 3 K warming is only about a 1% increase in the globally averaged surface temperature. To repeat, a 1% temperature change gives a 25% water vapor change! One important consequence is the strongly positive water vapor feedback, which was discussed in Chapter 5.

A 25% increase in the water vapor available in a column of air means that a thunderstorm can (potentially) produce 25% more rainfall. For this reason, it is expected that extreme local rainfall events will be *more* extreme in a warmer climate. The *globally averaged* precipitation rate is also expected to increase as the Earth warms, but at a much slower rate. As discussed earlier in this chapter, when the precipitation rate increases, the globally averaged radiative cooling of the atmosphere must also increase in order to maintain the energy balances of the atmosphere and the surface. Two factors favor an increased radiative cooling of the atmosphere:

Warmer air temperatures, which directly favor stronger emission

Increased concentrations of water vapor and CO_2, which are infrared emitters

Increased downward emission of infrared cools the atmosphere and warms the surface. The stronger atmospheric cooling "demands" more latent heat release, while the increased surface warming "enables" more surface evaporation. The hydrologic cycle speeds up accordingly.

Detailed calculations suggest that the precipitation rate will increase by only about 1% or 2% per kelvin, much less than the fractional increase in water vapor. Betts and Ridgway (1989), Held and Soden (2006), and Vecchi and Soden (2007) infer that *this implies a decrease in the convective mass flux, and also a slowing of the large-scale atmospheric circulation, as the climate warms.* The argument goes like this: The precipitation rate, P, can be expressed by $P = qM_c$, where q is a typical lower tropospheric water vapor concentration, and M_c is a convective mass flux. It follows that

$$\frac{dM_c}{M_c} = \frac{dP}{P} - \frac{dq}{q}.$$

(6.3)

Equation (6.3) shows that, if the fractional change in q is large, and the fractional change in P is less, then M_c must decrease in a warmer climate.

Now look back at Equation (6.2). It tells us that, in the descending branch of the Hadley-Walker circulation, the warming due to sinking must balance the radiative cooling. How will the various quantities that appear in Equation (6.2) change as the climate warms? Mimicking the approach used in Equation (6.3) above, we can write

$$\frac{d(\rho w)}{\rho w} = \frac{dQ}{Q} - \frac{d(\partial s/\partial z)}{\partial s/\partial z}. \tag{6.4}$$

We have already seen that the radiative cooling rate becomes stronger, which means that $\frac{dQ}{Q} > 0$ so that the subsidence-induced warming must increase too. Recall from Chapters 3 and 5, however, that the moist adiabatic lapse rate decreases as the temperature warms. This has major consequences. It means that $\partial s/\partial z$ becomes larger, so that $\frac{d(\partial s/\partial z)}{\partial s/\partial z} > 0$. Current research suggests that the increase in $\partial s/\partial z$ dominates, so that the right-hand side of (6.4) is negative. It follows from (6.4) that $\frac{d(\rho w)}{\rho w} < 0$, and so the downward mass flux of the Hadley-Walker circulation weakens. This implies that *the overall strength of the Hadley-Walker circulation decreases in a warmer climate*. The current generation of climate models predicts such changes by the end of the 21st century.

What about midlatitude winter storms? Recall from Chapter 5 that, because of the ice albedo feedback, the poles are expected to warm much more than the tropics. This means that the pole-to-Equator temperature gradient will become weaker. That gradient is the key to winter storms, so those storms are also expected to weaken in a warmer climate.

NEXT

In this and some of the preceding chapters, we have discussed possible future changes in atmospheric processes. Are such long-range predictions really possible? How are they different from weather forecasts? And speaking of weather forecasts, how are they made? What are the limits of the atmosphere's predictability, for both weather and climate? These and related issues are discussed in the next chapter.

7 PREDICTABILITY OF WEATHER AND CLIMATE

....

HOW FORECASTS ARE MADE

WEATHER FORECASTS ARE ALL AROUND US, ON TELEVI-sion, the radio, and the web. Where do they come from? In today's world, forecasts are made by solving equations that predict how the weather will change, starting from its observed current state. The equations comprise "mathematical models" that can be solved only by using very powerful computers. The models predict the winds, temperature, humidity, and many other quantities. The first such models were created in the 1950s. Through intensive worldwide efforts, driven partly by friendly competition among forecast centers, the forecast models have been refined, year by year, and have now reached an advanced state of development.

In practical terms, a forecast model is a large computer program, containing about a million lines of code.[1] The formulation of a forecast model is based on physical principles, such as conservation of mass, momentum, and energy. The spatial structure of the atmosphere is represented in a model by using grid points that, in the most advanced models, are now (2011) on the order of 30 km apart in the horizontal and 250 m apart in the vertical. The grid spacing of the model is often referred to

....

as its "resolution." A high-resolution model is one with many closely spaced grid points. Higher resolution requires a more powerful computer. As the years go by and computers improve, more grid points are added to the models, and we say that their resolution increases.

The global atmospheric models used to simulate climate change are essentially the same as the global forecast models, except that the climate models use lower resolution so that they can run faster. Today, a typical global atmospheric model used for climate simulation has grid points about 100 km apart in the horizontal and 500 m apart in the vertical.[2]

Important physical processes that are too small to "see" on a model's grid, like thunderstorms, are included by using statistical theories called "parameterizations." The parameterizations represent the effects of radiation, turbulence, and cloud formation. They are physically based theories that predict statistics. For example, a cloud parameterization determines the collective effect of many clouds, but does not represent individual clouds.

A forecast starts from the observed weather situation at a given moment in time. The gridded data used for this are called the "initial conditions." The data needed to create the initial conditions are obtained from a huge international observing network that includes surface stations, weather balloons, aircraft, and satellites, all of which are developed and operated under government funding. The data cannot be directly acquired on the regular grid used by a model, because weather stations, weather balloons, and satellite overpasses are irregularly situated in both space and time. The data are electronically gathered and gridded at the operational forecast centers, which also develop and run the forecast models. The models themselves are used to combine the many different types of data, collected at inconvenient places and times, into a coherent, self-consistent, quality-controlled, and gridded snapshot of the atmosphere that can be used as the initial conditions for a forecast. This very complex process is called "data assimilation."

The observing systems and forecast centers are operated by wealthy, technologically advanced nations and consortiums of nations. They are highly organized, quasi-industrial enterprises and adhere to strict and challenging schedules, so that the models can be run and forecasts can be delivered to the public every few hours.

When a forecast model starts up, it reads in the initial conditions and then predicts how the atmosphere will change over a short "time step" of a few minutes. Then it takes another time step, and another, until the forecast has advanced over as much simulated time as required by the human forecasters—perhaps 10 simulated days. The results are saved frequently as the model steps forward to create a record of the evolving state of the atmosphere over the period of the forecast. For reasons discussed later, it is now standard practice to run each forecast dozens of times. The simulations have to finish quickly enough to fit the schedule for releasing new forecasts.

When the computer-generated forecast is ready, it is analyzed by human forecasters at the national forecast centers, at local weather service offices, and sometimes also at media outlets. The humans use their judgment and experience to add value. Although human forecasters sometimes condescendingly refer to the model-generated forecast as the "numerical guidance," it is fair to say that, with rare exceptions, the forecast information that is provided to the public comes primarily from the forecast models.

Weather forecasting is a tough business. Forecasts are methodically compared with what really happens, and errors are quantified through detailed analyses. This is especially true for the attention-getting forecasts that are badly wrong. Over the months and years, statistics are compiled to evaluate a model's forecast skill. Forecast centers from around the world compete with each other, and over the past few decades the performance of the models has improved gradually but impressively. The improvements come from better data that can be used to make better initial conditions, and also from major improvements to the models themselves. Model improvements are achieved partly by adding more grid points and also, importantly, by improving the equations that are used to represent the physics of the atmosphere.

HOW PREDICTABLE IS THE WEATHER?

Are there limits to how good the forecasts can become? It turns out that, for fundamental physical reasons that are unrelated to any specific forecast method, it is impossible to make a skillful weather forecast more than about two weeks ahead.

As a first step toward understanding why, it is useful to consider three broadly defined sources of forecast error. First of all, there are errors in the initial conditions. The instruments (e.g., thermometers) used to make the various measurements are not perfect. To make matters worse, there are spatial and temporal gaps in the data, even in this era of weather satellites. The larger spatial scales tend to be relatively well observed because they cover a larger amount of real estate and so are sampled by more instruments. Smaller scales are less well sampled, and many small-scale weather systems are not observed at all. As a result, *the most serious errors in the initial conditions tend to be on the smallest spatial scales that can be represented on a model's grid*. We will come back to this point later. As the years go by, the observing system improves, and so the quality and quantity of the data that go into the initial conditions get better and better. Over the past 40 years or so, the gradual introduction of satellite data has enabled dramatic improvements in the accuracy of weather forecasts, especially for the Southern Hemisphere.

The second source of forecast errors is imperfections of the forecast models themselves. The resolution of a model is limited by the available computer power. In addition, the equations of the model are not exact, especially for the parameterized processes like radiation, turbulence, and cloud formation. The models are improving, year by year, but they will never be perfect.

Finally, and most importantly, our ability to predict the weather is limited by properties of the atmosphere itself. These are the most fundamental limitations because they are not related to any particular forecast model or observing system. At this point, you may be thinking of the Uncertainty Principle of quantum mechanics, which forbids exact measurements of the state of a system. The Uncertainty Principle is not important for weather prediction because there is a much larger source of uncertainty in classical physics, also intrinsic to the physical nature of the atmosphere: it is sometimes called *sensitive dependence on initial conditions*.

Recall that, in the presence of instability, the small difference between two initial states of the atmosphere can amplify with time. Many kinds of instability are at work in the atmosphere, acting on virtually all spatial scales. Small-scale shearing instabilities act on scales of meters or less. Buoyancy-driven instabilities, including cumulus instability, occur primarily on scales of a few hundred meters to a few kilometers. Baroclinic instability occurs on scales of thousands of kilometers.

As discussed in Chapters 3 and 4, the various scales interact with each other. Small-scale weather systems, such as cumulus clouds, can modify the larger scales, such as baroclinic waves, and the latter in turn can strongly influence where and when the cumulus clouds grow.

The combination of instability and scale interactions leads to sensitive dependence on initial conditions. The effects can be seen in weather forecasts. An example is given in the form of the "spaghetti diagrams" shown in

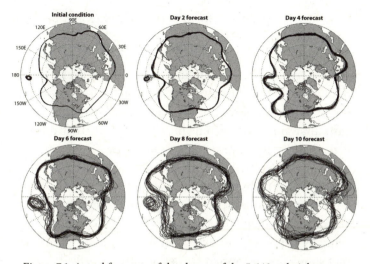

Figure 7.1. Actual forecasts of the shapes of the 5,640 m height contour for a pressure surface in the middle troposphere.

The plots look down on the North Pole, and North America is at the bottom center of each panel. The forecast times shown are the initial conditions (top left), and then, proceeding to the right across the upper row and then left to right across the bottom row, every two days after that, out to 10 days. The plotted curves show the positions of the contours as predicted in an ensemble of forecasts that are started from initial conditions that differ very slightly from each other. These are called "spaghetti diagrams."

Figure 7.1. The figure shows the results of actual forecasts. For simplicity, each highly detailed forecast is represented by a single curve that shows where the height of a pressure surface in the middle troposphere takes a certain value.[3] This is called a height contour. For each forecast, the winds would blow approximately parallel to the height contour shown, and primarily from west to east, weaving back and forth between higher and lower latitudes.

In the different panels of the figure, the forecast times shown are the initial conditions, and then every two days after that, out to ten days. For each forecast time, the plots show the results from an ensemble of 20 forecasts that are started from initial conditions that differ very slightly from those of the control run. There are 20 curves in each panel, one for each forecast. The 20 forecasts all start from almost but not quite the same initial conditions. You can tell that the initial conditions are very similar for all 20 forecasts because the contours essentially fall on top of each other in the initial condition plot shown in the top-left corner of the figure. By design, the differences in the initial conditions are so small that the accuracy of the observations is not sufficient to enable us to determine which one (if any) is "right," or closest to the true state of the atmosphere.

Two days into the forecast, the 20 predicted configurations of the weather are in pretty good agreement, except for a small region in the central Pacific, where the ensemble members differ considerably. The generally good agreement among the ensemble members gives confidence that the forecasts are reliable, and in fact the statistics show (and we know from common experience) that two-day forecasts are usually pretty good in 2011. After four days, the disagreements in the central Pacific have actually diminished, but the plots have become "fuzzy" virtually everywhere, because the different forecasts are beginning to disagree noticeably everywhere. By day 6, large disagreements have reappeared in the central Pacific, and the forecasts are also disagreeing quite a bit over northern Europe and eastern Asia. By day 10, the ensemble members have diverged so much that they provide little guidance on what the real atmosphere will do; they have become useless.

This example shows that after about a week, very small differences in the initial conditions can lead to large changes in forecasts performed with the same model. The details vary from case to case, and to some extent with season, but the basic pattern of gradually increasing forecast error is seen every time a forecast ensemble is run.

We understand why this happens. Our understanding comes largely from the work of Edward Lorenz, as first reported in an article titled "Deterministic Non-Periodic Flow," published in 1963. The title of Lorenz's article needs some explanation. A system is said to be deterministic if its future evolution is completely determined by a set of rules. The atmosphere obeys a set of rules that we call the laws of physics. It is, therefore, a deterministic system. A periodic system (or flow) is one whose behavior repeats exactly on some regular time interval. Despite the existence of the externally forced daily and seasonal cycles, the behavior of the atmosphere is nonperiodic; its previous history is not repeated. The predictability of periodic flows is a rather boring subject. If the behavior of the atmosphere were really periodic, the weather would certainly be predictable!

How does nonperiodic behavior arise? This is where scale interactions come in. The forcing of the atmosphere by the seasonal and diurnal cycles is at least approximately periodic. When scales do not interact, periodic forcing always leads to a response with the same period. When scales do interact, however, periodic forcing can lead to a nonperiodic response. *Nonperiodic behavior arises from scale interactions.*

Lorenz (1963) analyzed an idealized set of equations, loosely derived from a very simple atmosphere model. He found that, for some values of the parameters, all of the steady and periodic solutions are unstable. The instabilities arise on different scales, and the scales interact with each other. As a result, the model exhibits nonperiodic solutions; again, a periodic solution is, by definition, predictable.

The equations of the Lorenz's toy model are remarkably simple:

$$\dot{X} = -\sigma X + \sigma Y,$$
$$\dot{Y} = -XZ + rX - Y, \qquad\qquad (7.1)$$
$$\dot{Z} = XY - bZ.$$

The unknowns are X, Y, and Z. The dots on the left-hand sides of the three equations denote time derivatives, so the system consists of three coupled first-order ordinary differential equations. Initial conditions have to be supplied for the three unknowns.

The terms of (7.1) that involve products of the unknowns lead to interactions among time scales. Mathematically speaking, those terms are said to be "nonlinear," because the product of two unknowns is a nonlinear expression.

The model itself, as represented by the three equations in (7.1), is also said to be nonlinear. A nonlinear model permits scale interactions. Realistic models of the atmosphere, such as forecast models, are highly nonlinear.

The parameters σ, b, and r that appear in (7.1) are specified before the model is run. The numerical values $\sigma = 10$, $b = \frac{8}{3}$, and $r = 24.74$ are particular choices (not unique ones) that lead to nonperiodic behavior, as shown in Figure 7.2. In the figure, the state of the model is plotted in the (X,Y), (X,Z), and (Y,Z) planes. The curves show a time history. The system traces out a path that looks like two butterfly wings. The wings are called "attractors" because the solution tends to stay close to them, while the rest of the domain is avoided. At a randomly chosen time, the probability of finding the solution near one of the attractors is high. Occasionally the solution jumps from one attractor to the other. These "transitions" between attractors occur somewhat randomly. Lorenz demonstrated that the simple system shown in Equation (7.1) and Figure 7.2 exhibits sensitive dependence on initial conditions, analogous to the behavior of the forecasts shown in Figure 7.1. Lorenz's butterfly model, given by (7.1), illustrates that *even a simple system that includes scale interactions can be unpredictable*. In other words, unpredictable, complex behavior is not necessarily due to complexity in the definition of the system itself.

Every moment in time is the "initial condition" for the next moment. Systems that are sensitive to their initial conditions can be more generally described as sensitive to the details of their past histories. Such systems

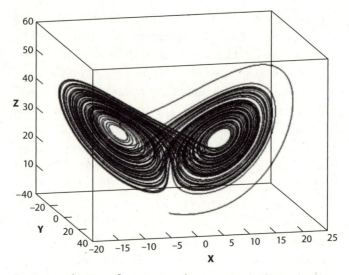

Figure 7.2. The Butterfly Attractor of Lorenz (1963), obtained as the solution of (1).

are said to exhibit "chaotic" behavior. The sensitive dependence of the state of the atmosphere on its past history famously suggests that the flap of a butterfly's wings in China could noticeably change the weather in North America a few days later. This is a second reason for calling the object shown in Figure 7.1 the Butterfly Attractor.

Lorenz was not the first to comprehend that there are fundamental limits to deterministic predictability. Poincaré (1912) recognized sensitive dependence on past history, and even discussed its relevance to weather forecasting. James Clerk Maxwell was also aware, during the

19th century, that as a result of instabilities, deterministic physical laws do not necessarily permit deterministic predictions at extended range (Harman, 1998, pp. 206–8). Lorenz's insight was that sensitive dependence on past history can occur even in very simple systems. He also emphasized the importance of scale interactions, in addition to instability.

It is the combination of instability and scale interactions that limits our ability to make skillful forecasts of the largest scales of motion. This idea is illustrated in Figure 7.3. Both instability and scale interactions are properties of the atmosphere itself; we cannot make them "go away" by improving our models or our observing systems. It is the properties of the atmosphere itself that lead to an intrinsic limit of deterministic predictability. We can say that sensitive dependence on past history imposes a "predictability time" or "predictability limit," beyond which the weather is unpredictable *in principle*. This limit is a property of the atmosphere itself, and for that reason there is no way around it. No matter what method is used to make a forecast, the predictability limit comes into play. Sensitivity to initial conditions is not a limitation that is tied to numerical weather prediction or any other specific forecast method. It applies to all forecast methods because it is a property of the atmosphere.

Errors on smaller scales double (grow proportionately) faster than errors on larger scales, simply because the intrinsic time scales of smaller-scale circulations are shorter. For example, the intrinsic time scale of a buoyant thermal in the boundary layer might be on the order of

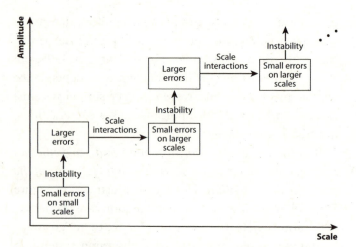

Figure 7.3. Sketch illustrating the roles of instability in leading to error growth, and of scale interactions in leading to the spreading of error from small scales to larger scales.

The figure could be continued indefinitely toward the upper right.

20 minutes, that of a thunderstorm circulation might be on the order of an hour, that of a winter storm might be two or three days, and that of a larger atmospheric wave that just fits on the Earth might be two weeks. Weather systems with longer intrinsic time scales are more predictable. As a result, the predictability limit is a function of scale; larger scales are generally more predictable than smaller scales.

When we reduce the errors in the initial conditions on small spatial scales by adding more observations, the range of our skillful forecasts is increased by a time increment that is approximately equal to the predictability time of the newly resolved small scales. For example, suppose that we are able to observe weather systems with intrinsic time scales of two days or longer and that we have models able to "resolve" such weather systems. By spending a lot of money, we can improve the observing system and the forecast model to permit prediction of weather systems with shorter intrinsic time scales—say, one day. This will lead to a one-day improvement in our forecast skill. Pushing the initial error down to smaller and smaller spatial scales is, therefore, a strategy for forecast improvement, but it is subject to a law of diminishing returns. Each successive refinement of the observing system and the models gives a smaller increase in the predictability time. At some point, we may choose to forego the improvement that would come from more accurate initial conditions and higher resolution models on the grounds that the benefit would be too small to justify the cost.

As discussed below, estimates show that small errors on the smallest spatial scales can grow in both amplitude and scale until they significantly contaminate the largest scales (comparable to the radius of the Earth) in about two to three weeks. Some aspects of atmospheric behavior may nevertheless be predictable on longer time scales. This is particularly true if they are forced by slowly changing external influences. An obvious example is the seasonal cycle. Another example is anomalies of the average weather that are associated with long-lasting sea surface temperature fluctuations, such as those due to El Niño. This point will be discussed further, later in this chapter.

We first discussed turbulence in Chapter 3. At this point, we are finally in a position to offer a definition of turbulence: *A flow is turbulent if its predictability time is shorter than the time scale of interest.* For example, baroclinic storms in midlatitude winter can be considered as turbulent eddies if we are interested in seasonal time scales, but they behave as highly predictable, orderly circulations if we are doing a one-day forecast. With this definition, a flow is turbulent or not depending on the time scale that we are paying attention to. To some extent, turbulence is in the eye of the beholder.

As an interesting analogy, the motions of the planets of our Solar System, and of the Earth's Moon, are well known to be highly predictable on time scales of thousands of years. This is why eclipses can be predicted very precisely many years ahead of time and why spacecraft can navigate accurately across the Solar System on trajectories that are computed years in advance. Nevertheless, the Solar System is known to be chaotic on longer time scales (e.g., Baytgin and Laughlin, 2008; Laskar, 1994; Laskar and Gastineau, 2009). If the time scale of interest is a few million years or less, the motions of the planets are analogous to highly predictable, laminar flow. If the time scale of interest is hundreds of millions of years or longer, the motions of the planets are analogous to turbulent flow.

QUANTIFYING THE LIMITS OF PREDICTABILITY

Lorenz (1969) outlined three distinct approaches to determine the limits of predictability. The first is based on entirely on models, the second entirely on observations, and the third uses both models and observations.

The Dynamical Approach

The first method is called the dynamical approach. A model is run two or more times, starting from similar but not quite identical initial conditions. The gradually increasing separation of the solutions is interpreted as "forecast error." The growth of the error can be studied, and experiments can be performed to see how the predictability of the model varies, for example with increasing resolution.

A problem with this approach is that it focuses on the predictability of a model, rather than the predictability of the real atmosphere. On the other hand, this type of calculation is easy to do, and the model is a perfect model of itself, because if we run the model twice with exactly the same initial conditions, we get exactly the same results.

Studies using the dynamical approach suggest that the doubling time for small errors with spatial scales of a few hundred km is about 5 days, and that the limit of predictability for the largest scales that fit on the Earth is about two weeks.

One of the earliest examples of the dynamical approach is the study of Charney et al. (1966). They used several global atmospheric models to study the growth of small perturbations. Figure 7.4 shows some of their results. The root-mean-square temperature error grows in both hemispheres, but more rapidly in the winter hemisphere, where

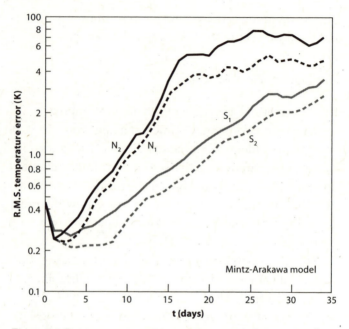

Figure 7.4. Root mean square temperature error in January simulations performed with a two-level global atmospheric model developed by A. Arakawa and Y. Mintz at the University of California, Los Angeles.

N and S denote the Northern and Southern Hemispheres, respectively. The subscripts 1 and 2 denote the two model levels.
Source: Charney et al. (1966).

the circulation is more unstable. The error does not continue to grow indefinitely; the growth stops at the point where the "forecast" is no better than a guess. Such a guess might take the form of a weather map pulled at random out of a huge cabinet full of many such maps. When the error has stopped growing, we say that it has "saturated."

The Empirical Approach

In the empirical approach, the atmosphere itself is used to predict the atmosphere. The idea is to search the weather records to find two states of the atmosphere, well separated in time, that are very similar to each other. Then we watch how the states become increasingly different on later days. Lorenz (1969) tried this. He examined the observational records and chose December 30, 1963, and January 13, 1965, as the best available analogs within a five-year record. I have looked at these weather maps, and they really don't look very similar to each other; as mentioned earlier, the weather does not repeat itself.

The results of the empirical approach show that the limit of predictability for the largest scales is about two weeks. They are thus consistent with the results of the dynamical approach.

Among the problems with the empirical approach are (a) it is hard to find "good" analogs, because the smallest initial "error" is not very small; (b) we cannot experiment with the initial error, because we have to take what nature gives us; and (c) the data cannot be used to study the growth of errors on very small scales, in part because such scales are not adequately observed.

The Dynamical-Empirical Approach

The "dynamical-empirical" approach uses both models and data. It can be based, for example, on the study of many weather forecasts, as compiled in the archives of a forecast center. The details will not be presented here, but the dynamical-empirical approach also supports the conclusion that the weather is predictable for only about two weeks.

The good news is that there is still room to improve today's operational forecasts. The bad news is that the improvement cannot continue indefinitely.

CLIMATE PREDICTION

If weather prediction is impossible beyond two weeks, how can climate prediction be contemplated at all? Two factors have the potential to make climate change prediction possible. First, the climate system has components with very long memories, including especially the ocean. Second, the climate system responds in systematic and predictable ways to changes in the external forcing.

The predictable response to external forcing means that it is possible to make a prediction without solving an initial-value problem! A good example is the seasonal cycle. It is possible to predict systematic differences in weather between summer and winter weather in any given location, such as Kansas City. We have a very good understanding of why those differences occur, in terms of the movement of the Earth in its orbit around the Sun. A climate model started on January 1 from initial conditions for the atmosphere, ocean, and land surface can predict, with ease, that the average July temperatures in Kansas City will be much warmer than those of the initial condition. In fact, it can predict that every single day in July will be warmer than January 1. That same model, started from the same January 1 initial condition, cannot predict the weather in Kansas City on January 15.

The movement of the Earth around the Sun is predictable. It represents a strong change in the external forcing of the climate system. The response of the climate system to that change in forcing is also predictable.

Climate prediction is possible when there is a strong, predictable change in the external forcing of the system. The ice ages are the predictable response to gradual changes in the Earth's orbital parameters. Similarly, the ongoing warming of the climate is a predictable response to the change in the composition of the atmosphere due to human activities.

Weather prediction is very different from climate prediction because changes in the day-to-day weather are not due to changes in the external forcing, while changes in climate are. Weather prediction is limited by sensitive dependence on past history. Climate prediction is not.

PUSHING THE ATTRACTORS AROUND

Palmer (1993, 1999) used Lorenz's butterfly model to illustrate the fact that chaotic systems (such as the atmosphere) respond in predictable ways to changes in external forcing. He modified the model to use the following equations, in place of the original equations given by (7.1):

$$\dot{X} = -\sigma X + \sigma Y + f_0 \cos\theta,$$
$$\dot{Y} = -XZ + rX - Y + f_0 \sin\theta, \qquad\qquad (7.2)$$
$$\dot{Z} = XY - bZ.$$

Here f_0 is an "external forcing" that tries to push X and Y in the direction of the angle θ in the (X,Y) plane. If we put $f_0 = 0$, Equation (7.2) reduces to (7.1). Figure 7.5 shows how different choices of θ affect the probability density function (PDF) of the solution in the (X,Y) plane. The maximums of the PDF are the attractors of the model. As θ changes, the locations of the maximums of the PDFs do not change much. This means that the attractors of the model are insensitive to θ. Note, however, that the maximums become stronger or weaker as θ changes. This means that by varying the forcing we can influence the relative amounts of time spent near each attractor. Also notice that there is some tendency for the attractors to be displaced in the direction in which the forcing acts.

Palmer's study shows that even a chaotic system responds in a statistically predictable way to sufficiently strong external forcing; this is why summers are predictably warm and winters are predictably cold. We can predict the response of the climate system to a sufficiently strong external forcing, provided that the forcing itself is predictable.

When the atmosphere interacts with the ocean and the land surface, the predictability of the combined system can be different from the predictability of the atmosphere alone. Such interactions are the subject of the next and penultimate chapter of this book.

....

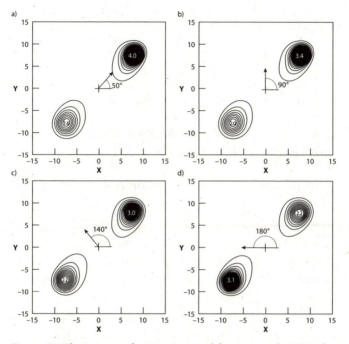

Figure 7.5. The impacts of various imposed forcings on the PDF of the butterfly model, as seen in the (X,Y) plane.

The plot shows the frequency with which the solution of the model visits each point in the plane. The arrows show the direction of the forcing, as represented by the angle θ in Equation (2). The oval-shaped contours denote regions that are visited frequently. These are the "attractors" of the model. Although the centroids of the attractors do not move as the angle of the forcing changes, the amount of time spent near each attractor does change. This can be seen from the "blackness" of the attractor maxima in the plot.

Source: Palmer (1999).

....

THE ATMOSPHERE AS A COMPONENT
OF THE LARGER CLIMATE SYSTEM

THIS BOOK IS ABOUT THE ROLE OF THE ATMOSPHERE IN climate, but the atmosphere strongly interacts with the ocean, the land surface, and the continental ice sheets. We say that the atmosphere is *coupled* to those other components of the system. To understand the climate system, we have to understand how its components interact with each other.

For example, as will be discussed below, the pattern of SST (sea surface temperature) is strongly influenced by the near-surface winds, but the winds, in turn, are influenced by the SST. The powerful SST changes associated with El Niño can be understood as one aspect of a coupled fluctuation of the tropical atmosphere and ocean, working together as a system. Over land, droughts, by definition, are prolonged periods with little or no rain, but the resulting dryness of the land surface can feed back to reinforce the dry weather pattern.

The atmosphere "feels" changes in the lower boundary most strongly in the form of variations in the surface temperature.

....

CHANGING THE TEMPERATURE
OF THE EARTH'S SURFACE

The surface temperature varies rapidly with time over land. Typical day-night temperature differences of the land surface and the air near the surface are on the order of 15 K. In contrast, ocean surface temperatures rarely change by more than a couple of degrees between day and night. "Heat capacity" is a bit of jargon that can be used to discuss this difference between land and ocean. The heat capacity of the Earth's surface can be defined as the amount of energy needed to change the "skin temperature," T_S, by a given amount. The skin temperature is defined as the temperature of an equivalent blackbody that would emit infrared radiation at a rate equal to the actual infrared emission by the Earth's surface. The skin temperature is governed by the energy budget of the Earth's surface, which can be written as

$$C\frac{\partial T_S}{\partial t} = N_S + H. \tag{8.1}$$

Here C is the heat capacity, N_S is the *net* downward energy flux at the Earth's surface (due to the combination of solar radiation, infrared radiation, and the sensible and latent heat fluxes), and H represents horizontal redistribution below or "inside" the Earth's surface. Equation (8.1) can be applied to both land and ocean. For the land, H is practically zero, but it can be large for the oceans because of lateral energy transport by ocean currents. Because of ocean currents, the energy absorbed by the oceans in one place can be given back to the atmosphere in a different place.

The concept of heat capacity sounds simple but is actually somewhat subtle. The specific heat of a material, such as dry air or liquid water, is a "material property" that can be measured in the laboratory. For example, we have used the specific heat of air at constant pressure, c_p, in this book, and you may remember, from some introductory science class, that the specific heat of water was used, at one time, to define the calorie. The heat capacity of the Earth's surface is not simply determined by the specific heat of soil or ocean water, however, because the energy flowing into or out of the surface is stored in a reservoir of finite depth. *The heat capacity of the surface therefore depends in part on the efficiency with which energy can be transported down into the material below the surface.* The heat capacity can vary in both space and time. It even depends on the time scale over which the surface energy fluxes vary. For example, the heat capacity for day-night changes is smaller than the heat capacity for seasonal changes.

The oceans have very large heat capacity, which is another way of saying that they respond slowly to energy input or loss. Part of the explanation is that the specific heat of liquid water is large. About 4,200 joules is needed to warm 1 kg of water by one kelvin. A second important factor is that the energy that flows into the ocean is distributed over many meters of water. First of all, sunlight often penetrates tens of meters into the water column, which means that the energy that it carries is distributed over tens of thousands of kilograms of water, per square meter. When and where the water is particularly clear, the heat capacity is larger, all other things being equal. In addition, turbulent mixing in the upper ocean redistributes energy vertically, sometimes over 100 m or more. Stronger, deeper mixing favors a larger heat capacity.

In contrast, the land surface has a very limited heat capacity because sunlight does not penetrate into soil or rock, and there is no turbulent mixing to carry heat down away from the surface. Molecular heat conduction is very weak. On daily time scales, heat fluxes into or out of the ground change the temperature of a layer of soil just a few centimeters thick. Even on seasonal time scales, heat is added to or removed from just the top meter or so of soil. This is why the heat capacity of the land surface is small, compared to that of the ocean. The net surface energy flux averages to nearly zero over land, even in a time average over a single day. To understand why, note that with $C \to 0$ and $H = 0$ (appropriate for land), Equation (8.1) implies that $N_S \to 0$. Small heat capacity means that "there is no room" to store energy, so there can't be any energy flow in or out.

As an example, suppose that you try to put energy into the land surface by supplying strong sunshine. Because the heat capacity of the land surface is small, the surface temperature increases very quickly, to a value high enough so that the combination of the net upwelling infrared and the sensible and latent heat fluxes just balance the absorbed sunshine. The temperature adjusts to whatever value is needed to make $N_S \cong 0$; a daily mean value of N_S over land would typically be a few watts per square meter.

For the ocean, with much larger values of C, "there is plenty of room" to store energy, and the daily mean values of N_S at particular locations can sometimes be as large as hundreds of watts per square meter. The large heat capacity of the upper ocean allows enormous energy storage on longer, seasonal time scales. The ability of the ocean currents to transport energy horizontally further enables large local values of N_S.

As discussed in Chapter 2, the Earth is thought to be emitting OLR somewhat more slowly than it is absorbing solar radiation, so that energy is gradually accumulating inside the system. Where does the energy go? Most of it ends up in the oceans, because they can easily store it.[1] The heat content of the oceans can, of course, be measured, using thermometers. So, you would think, it should be possible to measure the heat content of the oceans at two times several decades apart and subtract to determine the change. This has been done. In fact, the heat content of the oceans is being monitored over time, as shown in Figure 8.1. The figure shows an increase in heat content of about 1.5×10^{23} joules over a period of 40 years (which is about 1.3 billion s). Normalizing this by the Earth's surface area of about 500 million square km, we find an area-averaged energy input of about 0.25 W m^{-2}. This is too small to detect directly with current satellite systems. See Levitus et al. (2009) and von Schuckmann et al. (2009) for more discussion.

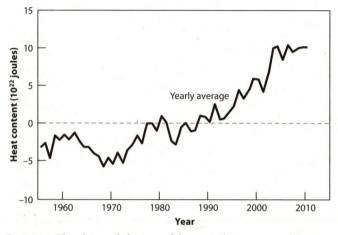

Figure 8.1. The observed changes of the ocean heat content, from 1955 to 2010.

The vertical axis is the change in the heat content of the ocean, in units of 10^{22} joules, relative to an arbitrary reference value (labeled as zero). The red curve shows averages over three months, and the red curve shows averages over a year.

Source: http://www.nodc.noaa.gov/OC5/3M_HEAT_CONTENT/.

INTERACTIONS OF THE ATMOSPHERE WITH THE OCEANS

The oceans cover about two-thirds of the Earth's surface. Their *average* depth is about 4 km. Water is heavy stuff; the mass of 1 m³ of water is 10^3 kg, which implies that a kilogram of water, spread over 1 m², is only 1 mm thick. The total mass of the oceans is about 1.3×10^{24} kg. For comparison, the mass of the atmosphere is about 250 times less, roughly 5×10^{21} kg.

Many properties of the atmosphere can affect the ocean below. These include the near-surface winds, which exert forces on the water; the near-surface temperature, which influences sensible heat exchange and downward longwave radiation; the near-surface humidity, which influences latent heat exchange and downward longwave radiation; and cloudiness, which can affect the surface solar and terrestrial radiation. Precipitation reduces the salinity of the surface waters, and evaporation increases it.

As discussed on the first pages of this book, the relatively large density of the air near the surface is due to the *compression* of the air by weight of the air above it. Does the same thing happen in the ocean? Not really. Although the density of the air can be changed fairly easily (try carefully squeezing a balloon filled with air), water is nearly incompressible (try squeezing a water balloon). The density of sea water is a complex but weak function of temperature, salinity, and pressure. The density of the water in the deepest parts of the ocean, 11 km below the surface, is only about 5% greater than the density near the surface! Because of the near-incompressibility of water, pressure effects (called "thermobaric" effects) are relatively unimportant; variations of the density are mainly due to changes in temperature and salinity. Warmer and fresher water is less dense and tends to float on top; colder and saltier water is more dense and tends to sink. Surface cooling and evaporation create dense water; surface heating and precipitation create light water. Note that the properties of the water are altered mainly near the surface; below the top hundred meters or so, the composition and temperature of water parcels remain nearly invariant, over decades and even centuries.

The property of the oceans that most directly affects the atmosphere is the SST, which influences the upward longwave radiation, the sensible heat flux, and the latent heat flux. Figure 8.2 shows the observed geographical distributions of the SST for January and July. Obviously, there is warm water in the tropics and colder water at higher latitudes. The bottom panel of the figure compares March, which is the end of the northern winter, to September, the beginning of the northern summer. The seasonal change of the SST is largest in the Northern Hemisphere, particularly on the western sides of the ocean basins, next to the east coasts of the continents. The largest seasonal differences in SST are about 15 K and are found off the east coasts of Asia and North America, where winter storms occasionally cause very cold air to pour out over the warm ocean water, progressively cooling the water as the winter unfolds. The depth to which this seasonal cooling penetrates is variable, but a ballpark figure is 100 m. The temperature of the water far below the surface undergoes virtually no seasonal change.

The warmest SSTs, at a particular latitude, are usually associated with tropical waters flowing poleward. The two best known examples are the Gulf Stream and the Kuroshio. East of North America, the Gulf Stream carries warm water from the tropics toward the Arctic. East

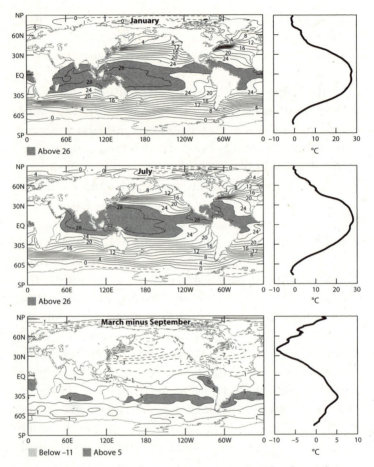

Figure 8.2. (a) SST distribution for January, in K. Values larger than 26°C are shaded. (b) Same for July. (c) The SST for March, minus the SST for September, in K.

In panel (c), values greater than 5 K are darkly shaded, and values less than −11 K are lightly shaded. Zonal means are shown on the right.

of Asia, the Kuroshio current does a similar job. There are also conspicuous cold currents near the west coasts of the continents. For example, the cold California Current is found west of North America, and the cold Humboldt Current is west of South America. Both the California and Humboldt Currents carry cold water from the polar regions toward the tropics.

Why are there warm currents on the western sides of the ocean basins, and cold currents on the eastern sides? Part of the answer has to do with the distribution of the near-surface atmospheric winds. Recall from Chapter 4 that the prevailing wind is from west to east in middle latitudes and from east to west in the tropics. When the wind blows over the ocean, surface drag pushes the water in the same direction that the wind is blowing. But then the Coriolis effect comes into play. In the Northern Hemisphere, the Coriolis acceleration causes the surface current to drift to the right of the surface wind, while in the Southern Hemisphere the current drifts to the left. The prevailing midlatitude westerlies push the ocean water toward the east, but the Coriolis effect induces an equatorward drift (again, to the right in the Northern Hemisphere, and to the left in the Southern Hemisphere). The tropical trade winds push the water toward the west, but the Coriolis effect then induces poleward drift. The wind stress pattern, combined with the Coriolis effect, causes the near-surface currents to form a pair of huge "gyres" in the Atlantic and Pacific basins, as shown in Figure 8.3. The gyres spin clockwise (as seen from above) in the Northern Hemisphere

CHAPTER 8

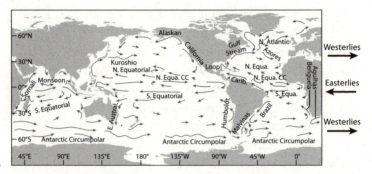

Figure 8.3. Map showing the surface ocean currents, which are largely driven by the winds.

The currents form loops, called "gyres." The gyres spin clockwise in the Northern Hemisphere, and counterclockwise in the Southern Hemisphere. The arrows on the right show the predominant wind directions, which are westerly in middle latitudes and easterly in the tropics.

Source: The underlying depiction of the ocean currents is from http://www.adp.noaa.gov/currents_map.html.

and counterclockwise in the Southern Hemisphere. The poleward branches of the gyres, such as the Gulf Stream and Kuroshio, occur on the western sides of the basins and tend to be warmer than average at a given latitude. The equatorward branches, such as the California Current and the Humboldt Current, occur on the eastern sides of the basins, and tend to be cooler than average at a given latitude.

There is more. Along the Equator, the east-to-west trade winds push the water toward the west, but the Coriolis effect causes the water to drift to the north in the Northern Hemisphere, and south in the Southern

Hemisphere. The surface waters thus spread out, away from the Equator, on both sides. This causes colder water to well up from below, leading to an SST minimum along the Equator, which can be seen in the zonal averages on the right of Figure 8.2. In the SST maps, the minimum is particularly noticeable in the eastern Pacific, west of South America. This westward flowing cold current is called the "cold tongue."

The winds also induce upwelling along coastlines. As discussed in Chapters 4 and 6, the low-level winds of the subtropical highs blow toward the Equator, along the west coasts of the continents, and then turn westward to join the trades. The equatorward winds drive equatorward coastal currents. The Coriolis effect causes the coastal currents to turn to the west, away from the coastlines. This induces upwelling along the coasts. For example, upwelling occurs along the west coast of South America, where the Humboldt Current is driven northward by the winds, and the Coriolis effect causes the water to drift toward the west, away from the coast. In a very similar way, upwelling also occurs in the California Current, which is driven southward by the winds of the California high.

As pointed out earlier, the coldness of these currents is not only due to upwelling. It is further enhanced by the fact that the currents flow from the directions of the poles. A third cause of the cold water is that, as discussed in Chapter 6, these regions are home to low-level stratus clouds, which reflect a lot of sunshine back to space. The occurrence of stratus clouds is favored by the coldness of the water and the sinking air of the subtropical highs. Finally, the coldness of the air above the cold water reinforces the subtropical highs. We thus have a beautifully symbiotic system, in which the SSTs, the winds, the upwelling, and the clouds all reinforce each other. Mutual reinforcement helps to explain why these subtropical climate regimes are observed to be very robust and persistent.

Once every few years, El Niño disrupts this idyllic picture.

ENSO

El Niño, La Niña, and the Southern Oscillation are dramatic fluctuations of the atmosphere and ocean together. They are beautifully described in the book of Philander (1990). During an El Niño, the eastern and central Pacific SST increases and the western Pacific SST decreases. A La Niña is like a mirror image of an El Niño. During a La Niña, the eastern Pacific SST becomes unusually cold and the western Pacific becomes unusually warm. El Niños are associated with weak upwelling west of South America, and La Niñas are associated with unusually strong upwelling. In some cases, the SST anomalies tend to move westward, as might be expected given the westward-moving currents driven by the westward-moving trade winds.

The Southern Oscillation is a systematic shifting of atmospheric mass, back and forth across the Pacific basin. Low atmospheric pressure and heavy rainfall follow the warm SSTs, and high atmospheric pressure and dry weather follow the cold SSTs. ENSO (which stands for "El Niño and the Southern Oscillation") is the name given to the collection of related phenomena that includes El Niño, La Niña, and the Southern Oscillation. ENSO is the best-understood example of a phenomenon that depends in an essential way on interactions between the atmosphere and the ocean.

El Niños occur, on the average, once every four years or so. We say that the "period" of ENSO is about four years, but in reality ENSO is only roughly periodic. Bjerknes (1969) achieved some important insights into how ENSO works.[2] He observed that a vigorous trade wind flow is encouraged by warm SSTs and low pressure in the west, accompanied by cold SSTs and high pressure in the east. At the same time, the trade winds induce upwelling west of South America and along the Equator. As the surface water flows westward across the vast Pacific basin, under the influence of the trades, it is gradually warmed by the Sun. When it arrives in the west, it forms the Warm Pool of the western Pacific. This is consistent with the canonical "Warm Pool–Cold Pool" scenario discussed in Chapter 6. La Niña can be interpreted as a particularly well developed case of the canonical scenario.

Bjerknes also pointed out, however, that if for any reason the trade winds were to "relax" (i.e., become weaker), the westward ocean currents would weaken and the SST would tend to increase in the eastern Pacific and decrease in the west, thus reinforcing the weakening of the trades. Such a situation could be described as an El

Niño. In short, Bjerknes recognized that the SST pattern and the trade winds are mutually reinforcing, for both El Niño and La Niña conditions.

Figure 8.4 shows a time series of SST and precipitation anomalies for a location in the tropical eastern Pacific Ocean with the peculiar name Niño 3.4. The anomalies are relative to an average (climatological) seasonal cycle. Rainfall increases during El Niños and decreases during La Niñas.

El Niños and La Niñas affect the weather not only in the tropics but also in the subtropics and middle latitudes. For example, an El Niño January tends to be unusually wet in California and across the southern part of North America and unusually dry over Australia. An El Niño July tends to be dry across southern Asia. We are gradually acquiring the ability to predict such weather anomalies, months in advance, using forecast models that couple the atmosphere and ocean together. It is not clear yet how predictable such "coupled ocean atmosphere weather" will prove to be.

After all the complexities of ocean currents, turbulent mixing, and so on, you might think that the interactions

Figure 8.4. Time series of SST anomalies (top panel) and precipitation anomalies (center panel) for a location in the eastern tropical Pacific Ocean, and how the SST and precipitation vary together (bottom panel).

The anomalies are defined relative to the climatological seasonal cycle. The precipitation data are not available before 1979.

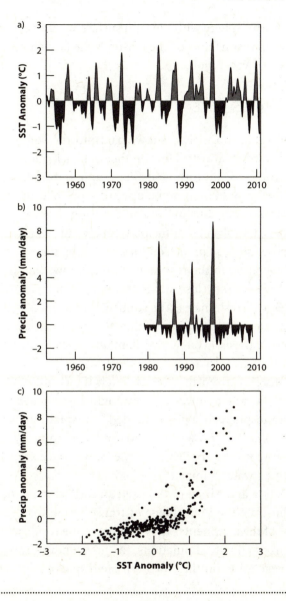

of the atmosphere with the land surface would be boring by comparison. Not so. The reason is the fascinating role played by vegetation.

STOMATES RULE

Rain water and melted snow soak into the soil, pulled downward by gravity. How do they get back out into the air? There are no turbulent fluxes underground, so the only way would seem to be upward molecular diffusion of water vapor. But molecular diffusion is a very slow process. If molecular diffusion were the only way out, the water would stay in the soil for a very long time.

Vegetation to the rescue. Everyone knows that plants need water. They lose water through openings in their leaves, called stomates. The stomates are needed to allow CO_2 to enter the leaves, so that the plant can perform photosynthesis. Water "accidentally" escapes through the stomates. To replace the lost water, plants pump moisture up from the soil, through their root systems. Then the water moves out, through the stomates, into the atmosphere, in a process called "transpiration." Roots and stomates allow water to move relatively quickly from the soil into the air. Plants on the land surface are solar-powered water pumps.

Shukla and Mintz (1982) used an early climate model to illustrate the importance of transpiration for the climate of the continents. They compared the results of two idealized climate simulations, for July conditions, when the vegetation on the Northern Hemisphere continents is active. In the first simulation, called the "wet soil case," they allowed moisture to evaporate from the soil that never dried out, in a crude imitation of "full-on" transpiration. In the second simulation, called the "dry soil case," no moisture at all was allowed to evaporate from the soil, regardless of how much rain fell. The dramatic differences between the two simulations are shown in Figure 8.5. The wet-soil results show plenty of precipitation over the continents, with cool temperatures. The dry-soil results show very little precipitation over the continents, with much warmer temperatures. These results illustrate, in a very dramatic way, whey climate models must include representations of the vegetation on the land surface.

Vegetation is important in other ways too. It affects the surface albedo. Very "rough" vegetation, like a forest, can significantly decrease the wind speed near the ground. Rain sometimes accumulates on the leaves of the plants and reevaporates from there, without ever touching the soil.

On time scales of days to weeks, rain or lack thereof can affect the "green-ness" and other characteristics of the vegetation. Dry weather causes the soil to dry out. When the soil is very dry, the plants try to protect themselves from desiccation by shutting down photosynthesis and closing their stomates. This cuts off transpiration. Warmer temperatures follow, which put additional stress on the plants. Such feedbacks can tend to perpetuate the dry weather and are believed to be important for the persistence of some droughts.

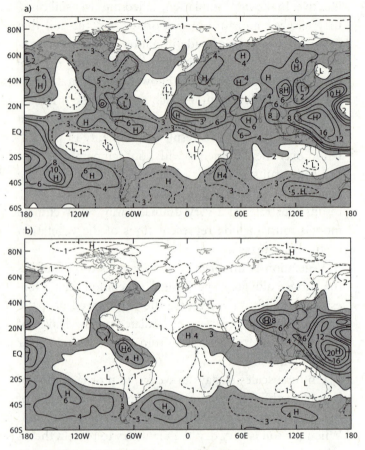

Figure 8.5. Left: Maps of the simulated precipitation.

Values larger than 2 mm day⁻¹ are shaded.

Right: Maps of the surface temperature.

In each case, the upper panel shows the results from the "wet soil" run and the lower panel those from the "dry soil" run.

Source: Shukla and Mintz (1982).

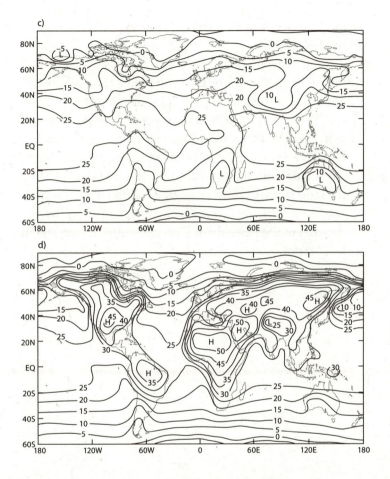

On time scales of years to decades, changes in climate can produce changes in the type and amount of vegetation, at a given location. For the reasons given above, such changes can feed back to influence climate change. This type of interaction between the atmosphere and the land surface is just beginning to be explored by climate modelers.

Both vegetation and soils are large reservoirs of carbon. Changes in these reservoirs have the potential to affect the atmospheric concentrations of both CO_2 and CH_4 (methane). If such changes occur in response to climate change, they can produce powerful feedbacks, but relatively slow ones that occur over decades or longer. Substantial amounts of carbon could be released from the land surface either through the elimination of tropical forests (by chainsaw or climate change) or by the thawing of the permafrost ("permanently" frozen soil in the Arctic).

READERS OF THIS BOOK ARE NO DOUBT AWARE THAT climate science is "hot," and will probably get even hotter in the years to come. The preceding chapters have lightly skimmed the surface of a few, selected topics. The atmospheric aspects of climate science are a lot broader, and much, much deeper than what I can cover here. I want to close the book by briefly mentioning a few particularly hot topics that were not discussed earlier. There are so many that it's hard to choose.

THE STRATOSPHERE

We have hardly talked about the stratosphere, but it is home to some amazing phenomena. For example, near the Equator, the stratospheric zonal winds undergo a wholesale reversal about once every two years. This is called the Quasi-Biennial Oscillation, or QBO (Baldwin et al., 2001). Westerlies are replaced by easterlies, which are then replaced by westerlies again. The QBO is like a huge Ferris wheel that periodically reverses its direction of rotation. The reversal is not confined to a few tropical locations; it happens all the way around the tropical belt. The basic mechanism of the QBO is now understood, although the details continue to be debated. The QBO is caused by waves that propagate up from the troposphere and exert forces on the stratospheric air.

"Sudden warmings" are also peculiar to the stratosphere. They occur near the winter pole, more often in the Northern Hemisphere. Most of the time the stratospheric air over the winter pole is very cold and surrounded by a ring of strong westerly winds, called the "polar night vortex." Several times in a typical winter, waves propagating up from the troposphere disrupt the polar night vortex. When that happens, much warmer air rapidly (over just a few days) replaces the cold air over the pole, and in extreme cases the polar night westerlies give way to easterlies. After a few more days, the polar night vortex is reestablished. Recent work shows that the stratosphere can influence the near-surface weather and climate (Baldwin et al., 2003). Sudden warmings play an important role in such influences.

Although increasing CO_2 concentrations are expected to warm the troposphere, they very strongly cool the stratosphere, and such cooling is quite apparent in the observations. Everyone has heard of the ozone hole, which is found in the stratosphere. The ozone-destroying chemical reactions occur on the ice particles that make up stratospheric clouds. Those clouds form only at very cold temperatures. The cooling of the stratosphere by increasing CO_2 may be promoting the formation of more polar stratospheric clouds.

THE MJO

Atmospheric scientists are struggling to understand a very large and powerful tropical weather system called the Madden-Julian Oscillation, or MJO (Zhang, 2005). The MJO occurs mainly over the remote tropical oceans and was not discovered until the early 1970s. It strongly influences precipitation over southern Asia and northern Australia, affecting the lives of literally billions of people. It is also believed to influence the timing and intensity of El Niños. Despite its importance, the MJO is perhaps the last type of weather system for which the basic physical mechanisms are not well understood.

AEROSOLS

Aerosols were briefly mentioned in Chapters 1 and 2. Many kinds of aerosols scatter solar radiation back to space, and thus tend to cool the planet. The data show a decrease in solar radiation reaching the surface during the 20th century, but this "global dimming" has recently weakened. The dimming is thought to have been caused by increasing aerosols associated with air pollution, and its recent cessation may be due to improvements in air quality. Aerosols formed from soot are dark in color, and so they absorb solar radiation. The absorption warms the atmosphere but prevents some solar radiation from reaching the surface, and so it cools the surface. A warming of the atmosphere due to solar absorption by soot can reduce the speed of the hydrologic cycle because it reduces the net radiative cooling of the atmosphere. The strong connections between the energy and water cycles were discussed in Chapter 6.

Finally, aerosols can affect the distribution and optical properties of clouds. Since cloud drops nucleate on aerosol particles, an increase in the number of aerosol particles favors a larger number of smaller cloud drops, for a given amount of liquid water. This makes the clouds brighter and so tends to increase the Earth's albedo. It has also been suggested that changes in aerosol abundance can affect precipitation.

VERY-HIGH-RESOLUTION CLIMATE MODELS

Climate models were briefly discussed in Chapter 7. Recent increases in computer power are making it possible to simulate the global atmosphere with grid cells just a few kilometers wide. This is important because it means that, for the first time, the models can actually simulate the growth and decay of individual large clouds, and other important but small-scale weather phenomena, some of which are associated with mountain ranges. Although such "global cloud-resolving models" can't yet be used for century-long climate simulations, further increases in computer power may make this possible within 10 years or so. The result will be a revolution in our ability to simulate the climate.

....

THE CLIMATES OF THE PLANETS

We are beginning to understand the climates of the other planets in our Solar System, which have both similarities and differences from the Earth's. This subject is discussed at length elsewhere, so I will limit myself here to a few brief comments.

Mars has a very thin atmosphere that is composed almost entirely of CO_2. As mentioned earlier, the polar regions of Mars get so cold in winter that the atmosphere actually condenses onto the surface, in the form of dry ice. It returns to the atmosphere in the spring. The Martian atmosphere is very dry, but there are a few water clouds. Global dust storms are very important to the climate of Mars. Recall that the obliquity is the angle that a planet's axis of rotation makes with the plane of its orbit. As mentioned in Chapter 2, the obliquity of Mars changes a lot on geologic time scales, and this is believed to have produced major changes in the Martian climate.

The very massive atmosphere of Venus is also composed almost entirely of CO_2. Venus is very bright, when seen from Earth, because it is shrouded by clouds. The clouds reflect so much sunlight that the solar radiation absorbed by Venus is actually less than that absorbed by the Earth, even though Venus is considerably closer to the Sun. Despite its high planetary albedo, the greenhouse effect of Venus's thick CO_2 atmosphere makes the surface very hot, on the order of 450°C. The pressure and temperature 50 km above the surface of Venus are comparable to those at sea level on Earth.

Gigantic Jupiter has one of the most interesting atmospheres in our Solar System. Jupiter rotates more than twice as fast as the Earth, and its atmosphere correspondingly has more "bands" than the Earth's. Most if not all of Jupiter is atmosphere, in contrast to the Earth, whose atmosphere is just a thin shell covering a rocky ball. Primordial energy is leaking out from deep inside Jupiter; the planet emits about 10% more radiation than it absorbs. The upwelling energy is carried in part by enormous thunderstorms, complete with lightning. The huge Jovian weather system that we call the Great Red Spot has persisted for hundreds of Earth years, raising questions about the predictability of Jupiter's weather.

Through studies of the other planets, climate science is being subjected to challenging observational tests that will lead to better understanding of our own planet. Over the coming decades, the study of the Earth's climate and the science of other planetary atmospheres may gradually blend together. Climate science is a big, beautiful, challenging, and important subject, of tremendous societal relevance. It is a worthy focus for a life's work.

CHAPTER 1

1. As words spoken by the fictional Lazarus Long, in the novel *Time Enough for Love*.

2. Latitude is an angular measure of position on the spherical Earth, increasing from south to north. By convention, the latitude of the Equator is zero, that of the South Pole is −90°, and that of the North Pole is +90°. Similarly, longitude is an angular measure of position, increasing from west to east. By convention, the longitude of the Greenwich Observatory, near London in the United Kingdom, is zero, and that of the Date Line is 180°. New York City is 74° of longitude west of Greenwich, and San Francisco is 122° of longitude west of Greenwich.

3. The concept of heat capacity will be explained at some length in Chapter 8.

4. Gases, including air, are fluids. A fluid is a material that yields freely to forces that try to change its shape. The same definition can be stated more technically this way: A fluid is a material that cannot be in equilibrium with a shear stress.

5. It is somewhat debatable whether all of these parameters are really external. The answer depends in part on the time span considered. For example, on geologic time scales precipitation erodes mountain ranges, and volcanoes can change the composition of the air. Therefore, on time scales of millions of years, the Earth's rocky parts have to be considered as internal components of the climate system.

6. Here the units are on the Kelvin scale (*not* "degrees kelvin"), an absolute temperature scale defined in such a way that

the freezing point of water is approximately 273 K. A kelvin is the same size as a degree Celsius, so the temperature in kelvin is the temperature in degrees Celsius, plus 273.

7. Here we neglect the weight of the air pressing down on the top of the water column.

8. Realistic variations of temperature with height are slow enough so that our simplifying assumption that the temperature is constant does not lead to large errors in this analysis.

9. Standard atmospheres have been defined for almost a hundred years. At first, they were used to calibrate aircraft altimeters and to calculate the trajectories of artillery shells. Now they serve mainly to introduce the vertical structure of the atmosphere in books like this one.

10. There is a weak chemical source of water vapor in the stratosphere due to the oxidation of methane.

11. Nitrogen does not condense under conditions found in the Earth's atmosphere, so the latent heat of nitrogen plays no role in the climate system. The atmosphere of Mars is composed mostly of carbon dioxide, and gaseous CO_2 is exchanged with solid CO_2 ("dry ice") near the Martian poles in winter. As a result, the mass of the Martian atmosphere undergoes large seasonal changes! If the Earth's atmosphere were much colder, something similar would happen with nitrogen.

CHAPTER 2

1. The other ways are the capture of interplanetary material (dust, meteors, etc.) and the action of the gravitational tides on the ocean, atmosphere, and solid Earth.

2. Unfortunately, only a small percentage of the energy used by a conventional light bulb actually appears as visible light; the rest is wasted by heating the bulb. Newer lighting technologies are more efficient.

3. A blackbody is an idealized object that absorbs all radiation impinging on it and emits radiation that varies with wavelength in a simple, smooth way that depends only on temperature.

4. The term "flux" is used to describe the flow of any quantity from place to place in the atmosphere, or in this case across a boundary of the atmosphere.

5. The term "greenhouse" is used only by way of analogy. As many people have pointed out, it is a poor analogy. A real greenhouse traps air that has been warmed by the Sun.

6. The even simpler case of a single atmospheric layer turns out to be uninteresting.

7. In fact, by "writing small," I was once able to squeeze the argument onto the back of a business card.

CHAPTER 3

1. This is the longest and most challenging chapter of the book. Take your time. The subjects covered in this chapter are very important for climate.

2. There is actually one more way that the atmosphere can exchange energy with the Earth's surface, namely the surface drag force. This is the frictional force that the near-surface air experiences. It generally depletes the energy of the atmosphere by doing work on the surface. For example, over the oceans, the surface drag pushes on the water, and by doing so transfers kinetic energy from the atmosphere to the ocean. This drives the ocean currents and also generates waves on the sea surface. Over land, the surface drag rustles the leaves, turns windmills, and sometimes raises dust. Averaged over the Earth, these "mechanical" flows of energy from the atmosphere to the Earth's surface are very small, on the order of 0.01 W m^{-2}. They are tiny compared to the radiative, latent, and sensible energy fluxes.

3. Of course, there has to be a compensating downward flow of air somewhere else.

4. The interiors of cumulus clouds (and practically all other clouds) are turbulent, but cloud-scale vertical motions produce vertical fluxes independently of this turbulence.

5. The same terminology is used by oceanographers and other geoscientists.

6. Unfortunately, the standard atmosphere is highly idealized and does not include a "mixed layer" of vertically uniform dry static energy near the surface. It would be more realistic if it did.

7. Physicists use the term "gravity wave" to refer to something completely different.

8. As mentioned in Chapter 1, the condensation process is mediated by the availability of aerosol particles that are chemically suitable for nucleating drops; these are called cloud condensation nuclei, or CCN. We will assume for simplicity that sufficient CCN are always available, and that condensation occurs as soon as saturation is exceeded. This is almost always true.

9. If the air becomes cold enough to permit ice clouds, the relative humidity can exceed 100% by a substantial margin.

10. The moist static energy as defined here is affected by phase changes involving ice, but we neglect those for simplicity. It is possible to define the moist static energy in a slightly more complicated way, such that phase changes involving ice do not affect it.

11. Here we are assuming that the parcel's buoyancy is determined entirely by its temperature. In reality, the parcel's buoyancy is also affected by its water vapor and liquid water contents, but we neglect those secondary effects here, for simplicity.

12. An additional condition is that the parcel must be able to reach its level of free convection.

13. As Sherlock Holmes said in *The Sign of the Four*, "When you have eliminated the impossible, whatever remains, however improbable, must be the truth."

14. The specific humidity and dry static energy are examples of quantities that are defined per unit mass.

15. These various assumptions can be exactly true, depending on the details of how the average is defined. For example, the average of a sum is exactly equal to the sum of the averages if the averaging operator is linear.

CHAPTER 4

1. A sidereal day is the time required for the Earth to complete one rotation on its axis. It is slightly shorter than a solar day because of the movement of the Earth in its orbit around the Sun.

2. You might be wondering why the moment arm in (1) is not the total distance from the Earth's axis of rotation to a parcel of air that might be far above the Earth's surface. Well, really it is, but the radius of the Earth gives a good approximation to the total distance, simply because the Earth's atmosphere is so thin compared to the Earth's radius. For a planet with a thicker atmosphere, such as Jupiter, this approximation would not be justified.

3. My apologies to the inhabitants of the Southern Hemisphere.

4. I am simplifying slightly here.

5. Here is the cumbersome terminology: "hurricane" (the north Atlantic Ocean, the northeast Pacific Ocean east of the Date Line, or the south Pacific Ocean east of 160°E), "typhoon" (the northwest Pacific Ocean west of the Date Line), "severe tropical cyclone" (the southwest Pacific Ocean west of 160°E or southeast Indian Ocean east of 90°E), "severe cyclonic storm" (the north Indian Ocean), and "tropical cyclone" (the southwest Indian Ocean). Source: http://www.aoml.noaa.gov/hrd/tcfaq/A1.html.

CHAPTER 7

1. For comparison, a typical computer operating system contains between 10 million and 100 million lines of code.

2. These are rough numbers.

3. Each forecast actually predicts the detailed evolution of the weather, including a very large number of parameters, but only a single height contour is plotted in these diagrams. The particular height contour is chosen based on a combination of experience and convention, and the details of the choice are not important for the present discussion.

CHAPTER 8

1. Additional energy is being used to melt snow and ice, for instance on the large continental ice sheets of Greenland and Antarctica.

2. This is the same man whose work on cumulus clouds was discussed in Chapter 3. His father, Vilhelm Bjerknes, was also a highly accomplished atmospheric scientist.

Glossary

Advection—Changes in the property of the air, at a fixed location, due to the movement of the air from other locations. For example, the wind may carry cooler air toward your location, in which case the temperature will tend to fall there. Advection can affect any property of the air that is "carried with" the air.

Aerosol—Tiny liquid and solid particles floating in the air, including dust, sea salt, and soot. Typical particles have diameters on the order of 10^{-7} m. Aerosol concentrations often reach hundreds of particles per cubic centimeter. Some aerosols serve as nuclei for the formation of cloud droplets or ice crystals.

Albedo—The reflected fraction of radiation, usually solar radiation. The Earth's albedo is the reflected solar energy divided by the incident solar energy. Its value is near 0.3.

Baroclinic wave—Winter storms that grow as warm air rises and flows poleward, while cooler air sinks and flows equatorward.

Boundary layer—The lowest portion of the troposphere, which is filled with turbulence.

Brightness temperature—The temperature of a black body that emits a given radiative flux. The brightness temperature of the Earth is about 255 K, which is the temperature of a black body that emits radiation equal to the Earth's outgoing longwave radiation.

CAPE—An acronym for "convective available potential energy," which is a measure of the positive buoyancy that can be generated by a cumulus updraft.

Chaos—The unpredictable behavior of a system that exhibits sensitive dependence on its initial conditions.

Climate—The average weather at a place, usually over a period of 30 years or more.

Conditional instability—The instability that leads to the growth of cumulus clouds. The "condition" is that sufficient moisture is available to allow rising air to reach its level of free convection.

Conservative variable—A property of the air that does not change as the air moves. There are no perfectly conservative variables, but some are more conservative than others. Examples of approximately conservative variables include the dry static energy, the moist static energy, and the angular momentum per unit mass.

Convection—Circulations driven by positive or negative buoyancy forces. The term is usually applied to small-scale circulations, including thermals in the boundary layer and cumulus clouds.

Cumulus clouds—Clouds that are produced by moist updrafts in which positive buoyancy is generated by latent heat release. See "cumulus convection."

Cumulus convection—The growth of clouds in the form of positively buoyant updrafts. The positive buoyancy is created by the release of latent heat as water vapor condenses in the updrafts.

Dry convection—Convection that does not involve phase changes of water.

Dry static energy—The sum of the enthalpy and the gravitational potential energy. The dry static energy is approximately conserved in the absence of heating.

Eccentricity—A quantitative measure of the extent to which the shape of the Earth's orbit differs from a perfect circle. For a circular orbit, the eccentricity would be zero. The eccentricity changes on geologic time scales.

El Niño—A warming of the sea surface temperature in the eastern and central Pacific Ocean, near the Equator.

ENSO—An acronym for "El Niño and the Southern Oscillation." The Southern Oscillation is a seesaw of surface pressure between the eastern and western tropical Pacific.

Environment—The air surrounding a cumulus cloud.

Equinox—Time when the Sun is directly over the Equator at noon and the Earth's axis of rotation points neither toward nor away from the Sun. In the present era, the equinoxes occur near September 22 and March 22, each year. The dates of the equinoxes change on geologic time scales.

Feedback—A causal "loop" in which the output of a process modifies the input. In a positive feedback the output strengthens the input, while in a negative feedback the output weakens the input.

Geostrophic balance—A balance between the Coriolis acceleration and the horizontal pressure-gradient force. Geostrophic balance is a useful approximation on large scales, away from the Equator.

Hadley cell—An large-scale overturning of the tropical troposphere, with a rising motion near the Equator and a sinking motion in the subtropics.

Hydrostatic balance—A balance between the weight of the air and the vertical pressure-gradient force. Hydrostatic balance is an excellent approximation on large scales.

Jet stream—A strong, river-like current of air. The most familiar jet streams blow from west to east and are found near the tropopause at about 30° north and south.

Kelvin—An absolute temperature scale in which the units, i.e., kelvins, are the same size as degrees Celsius.

La Niña—A cooling of the sea surface temperature in the eastern and central Pacific Ocean, near the Equator.

Lapse rate—The rate at which the temperature decreases with height.

Latent heat—The energy needed to evaporate a given mass of water. The same amount of energy is released in the form of enthalpy when the water vapor condenses.

Latent heat flux—The energy flow per unit time and per unit area due to the movement of water vapor, with its associated latent heat.

Latitude—An angular measure of position on the spherical Earth, increasing from south to north. By convention, the latitude of the Equator is zero, that of the South Pole is –90°, and that of the North Pole is +90°.

Longitude—An angular measure of position, increasing from west to east. By convention, the longitude of the Greenwich Observatory, near London in the United Kingdom, is zero, and that of the Date Line is 180°.

Lorenz butterfly model—A simple set of equations, discovered by Edward Lorenz, that produces a chaotic solution with an attractor in the shape of a butterfly.

Meridian—A line of constant longitude.

Meridional—The south-to-north component of a vector, e.g., the "meridional wind." A meridional average is an average in the north-south direction.

Moist static energy—The sum of the dry static energy and the latent heat of water vapor. The moist static energy is approximately conserved in the absence of heating, even when water vapor is condensing or liquid water is evaporating.

Obliquity—The angle that the plane of the Earth's Equator makes with the plane of the Earth's orbit.

Parcel—A point-like particle of air that moves with a well-defined velocity.

Predictability—The degree to which a process or phenomenon is predictable.

Predictability time—The time scale on which a given system is predictable. The system is not predictable on time scales longer than its predictability time.

Saturation vapor pressure—The partial pressure of water vapor at which the vapor is in equilibrium with a liquid water surface.

Saturation moist static energy—The moist static energy that would exist if the air were saturated with respect to water vapor, for a given dry static energy, temperature, and pressure.

Saturation specific humidity—The specific humidity that would exist if the air were saturated with respect to water vapor, at a given temperature and pressure.

Sensible heat flux—The energy flow per unit time and per unit area due to the movement of enthalpy.

Solstice—Times when the Sun's apparent position in the sky reaches its seasonal extremes, which are about 23° away from the Equator. In the present era, the solstices occur near December 22 and June 22, each year. The dates of the solstices change on geologic time scales.

Sounding—The vertical distribution of the properties of the air, e.g., a "temperature sounding."

Specific humidity—The mass fraction of water vapor in the air.

Stratiform clouds—Flat, neutrally buoyant clouds that often form from the outflows of cumulus clouds.

Stratopause—The upper limit of the stratosphere.

Stratosphere—A meteorologically quiet layer of air between (approximately) 12 km and 50 km above the surface, within which the temperature generally increases upward due to the absorption of solar ultraviolet radiation by ozone.

Terrestrial radiation—Infrared radiation that is emitted by the Earth.

Thermal wind—The rate of change of the horizontal wind with height.

Tropical cyclone—A hurricane or typhoon that forms over the tropical oceans and sometimes moves over land. Tropical cyclones are powered by energy flowing from the ocean to the atmosphere. The energy flow is promoted by the strong winds of the tropical cyclone.

Tropopause—The upper limit of the troposphere.

Troposphere—The meteorologically active layer of air between the surface and (approximately) 12 km high. Within the troposphere, the temperature generally decreases upward.

Turbulence—A turbulent flow is one for which the predictability time is shorter than the time scale of interest.

Zonal—The west-to-east component of a vector, e.g., the "zonal wind." A zonal average is an average in the east-west direction.

Suggestions for Further Reading

CHAPTER 1

Hartmann, D. L., 1994: *Global physical climatology*. International Geophysics Series, Vol. **56**. Academic Press, San Diego, 411 pp.

Imbrie, J., and K. P. Imbrie, 1986: *Ice ages: Solving the mystery*. Harvard University Press, Cambridge, Mass., 224 pp.

Lyden-Bell, R. M., S. C. Morris, J. D. Barrow, J. L. Finney, and C. Harper, Eds., 2010: *Water and life: The unique properties of H$_2$O*. CRC Press, New York, 396 pp.

Seinfeld, J. H., and S. N. Pandis, 1998: *Atmospheric chemistry and physics: From air pollution to climate change*. John Wiley, New York, 1,232 pp.

Trenberth, K. E., J. T. Fasullo, and J. Kiehl, 2009: Earth's global energy budget. *Bull. Amer. Meteor. Soc.*, **90**, 311–23.

CHAPTER 2

Barkstrom, B., E. F. Harrison, G. L. Smith, R. N. Green, J. Kibler, R. Cess, and the ERBE Science Team, 1989: Earth Radiation Budget Experiment (ERBE) archival and April 1985 results. *Bull. Amer. Meteor. Soc.*, **74**, 591–98.

Crowley, T. J., and G. R. North, 1991: *Paleoclimatology*. Oxford University Press, Oxford, 339 pp.

Fleming, J. R., 2009: *The Callendar effect*. American Meteorological Society, Boston, 176 pp.

Pierrehumbert, R. T., 2011: Infrared radiation and planetary temperature. *Phys. Today*, **64**, 33–38.

Solomon, S., D. Qin, M. Manning, Z. Chen, M. C. Marquis, K. B. Averyt, M. Tignor, and H. L. Miller, Eds., 2007: *Climate change 2007: The physical science basis. Contribution of Working Group I to the Fourth Assessment Report of the Intergovernmental Panel on Climate Change*. Cambridge University Press, Cambridge, UK, 996 pp.

Trenberth, K. E., J. T. Fasullo, and J. Kiehl, 2009: Earth's global energy budget. *Bull. Amer. Meteor. Soc.*, **90**, 311–23.

Weart, S. R., 2008: *The discovery of global warming*, rev. and exp. ed. Harvard University Press, Cambridge, Mass., 240 pp.

CHAPTER 3

Cripe, D. G., and D. A. Randall, 2001: Joint variations of temperature and water vapor over the midlatitude continents. *Geophys. Res. Lett.*, **28**, 2613–26.

Emanuel, K. A., 1994: *Atmospheric convection*. Oxford University Press, Oxford, 580 pp.

Randall, D. A., M. Khairoutdinov, A. Arakawa, and W. Grabowski, 2003: Breaking the cloud-parameterization deadlock. *Bull. Amer. Meteor. Soc.*, **84**, 1547–64.

CHAPTER 4

Emanuel, K. A., 2005: *Divine wind: The history and science of hurricanes*. Oxford University Press, Oxford, 296 pp.

Holton, J. R., 1992: *An introduction to dynamic meteorology*, 3rd ed. International Geophysics Series, Vol. **48**. Academic Press, San Diego, 511 pp.

Marshall, J., and R. A. Plumb, 2008: *Atmosphere, ocean, and climate dynamics*. Elsevier, Amsterdam, 219 pp.

CHAPTER 5

Bony, S., R. Colman, V. M. Kattsov, R. P. Allan, C. S. Bretherton, J.-L. Dufresne, A. Hall, S. Hallegatte, M. M. Holland, W. Ingram, D. A. Randall, B. J. Soden, G. Tselioudis, and M. J. Webb, 2006: How well do we understand and evaluate climate change feedback processes? *J. Climate*, **19**, 3445–82.

Cess, R. D., G. L. Potter, J. P. Blanchet, G. J. Boer, S. J. Ghan, J. T. Kiehl, H. Le Treut, Z.-X. Li, X.-Z. Liang, J. F. B. Mitchell, J.-J. Morcrette, D. A. Randall, M. Riches, E. Roeckner, U. Schlese, A. Slingo, K. E. Taylor, W. M. Washington, R. T. Wetherald, and I. Yagai, 1989: Interpretation of cloud-climate feedback as produced by 14 atmospheric general circulation models. *Science*, **245**, 513–16.

Held, I. M., and B. J. Soden, 2000: Water vapor feedback and global warming. *Ann. Rev. Energ. Environ.*, **25**, 441–75.

———, 2006: Robust responses of the hydrological cycle to global warming. *J. Climate*, **19**, 3354–60.

Lacis, A. A., G. A. Schmidt, D. Rind, and R. A. Ruedy, 2010: Atmospheric CO_2: Principal control knob governing Earth's temperature. *Science*, **330**, 356–59.

Solomon, S., D. Qin, M. Manning, Z. Chen, M. C. Marquis, K. B. Averyt, M. Tignor, and H. L. Miller, Eds., 2007: *Climate change 2007: The physical science basis. Contribution of Working Group I to the Fourth Assessment Report of the Intergovernmental Panel on Climate Change*. Cambridge University Press, Cambridge, UK, 996 pp.

CHAPTER 6

Held, I. M., and B. J. Soden, 2006: Robust responses of the hydrological cycle to global warming. *J. Climate*, **19**, 3354–60.

Philander, S. G., 1990: *El Niño, La Niña, and the Southern Oscillation*. Academic Press, New York, 293 pp.

Pierrehumbert, R. T., 1995: Thermostats, radiator fins, and the local runaway greenhouse. *J. Atmos. Sci.*, **52**, 1784–1806.

CHAPTER 7

Donner, L., W. H. Schubert, and R. C. J. Somerville, Eds., 2011: *The development of atmospheric general circulation models: Complexity, synthesis, and computation*. Cambridge University Press, Cambridge, UK, 272 pp.

Gleick, J., 1987: *Chaos: Making a new science*. Viking Press, New York, 352 pp.

Harman, P. M., 1998: *The natural philosophy of James Clerk Maxwell*. Cambridge University Press, Cambridge, UK, 232 pp.

Lorenz, E. N., 1969: Three approaches to atmospheric predictability. *Bull. Amer. Meteor. Soc.*, **50**, 345–49.

———, 1995: *The essence of chaos*. University of Washington Press, Seattle, 227 pp.

Palmer, T. N., 1993: Extended range atmospheric prediction and the Lorenz model. *Bull. Amer. Meteor. Soc.*, **74**, 49–65.

Woods, A., 2006: *Medium-range weather prediction: The European approach*. Springer, Berlin, 270 pp.

CHAPTER 8

Marshall, J., and R. A. Plumb, 2008: *Atmosphere, ocean, and climate dynamics*. Elsevier, Amsterdam, 219 pp.

Philander, S. G., 1990: *El Niño, La Niña, and the Southern Oscillation*. Academic Press, New York, 293 pp.

————, 2010: Tilting at connections, from Pole to Equator. *Science*, **328**, 1488–89.

CHAPTER 9

Baldwin, M. P., L. J. Gray, T. J. Dunkerton, K. Hamilton, P. H. Haynes, W. J. Randel, J. R. Holton, M. J. Alexander, I. Hirota, T. Horinouchi, D. B. A. Jones, J. S. Kinnersley, C. Marquardt, K. Sato, and M. Takahashi, 2001: The Quasi-Biennial Oscillation. *Rev. Geophys.*, **39**, 179–229.

Baldwin, M. P., D. W. J. Thompson, E. F. Shuckburgh, W. A. Norton, and N. P. Gillett, 2003: Weather from the stratosphere? *Science*, **301**, 317–19.

Pierrehumbert, R., 2010: *Principles of planetary climate*. Cambridge University Press, Cambridge, UK, 652 pp.

Wild, M., H. Gilgen, A. Roesch, A. Ohmura, C. N. Long, E. G. Dutton, B. Forgan, A. Kallis, V. Russak, and A. Tsvetkov, 2005: From dimming to brightening: Decadal changes in solar radiation at Earth's surface. *Science*, **308**, 847–50.

Zhang, C., 2005: The Madden-Julian Oscillation. *Rev. Geophys.*, **43**, RG2003.

Bibliography

Adler, R. F., G. J. Huffman, A. Chang, R. Ferraro, P. Xie, J. Janow-iak, B. Rudolf, U. Schneider, S. Curtis, D. Bolvin, A. Gruber, J. Susskind, and P. Arkin, 2003: The Version 2 Global Precipitation Climatology Project (GPCP) monthly precipitation analysis (1979–present). *J. Hydrometeor.*, **4**, 1147–67.

Arakawa, A., 1975: Modeling clouds and cloud processes for use in climate models. In *The physical basis of climate and climate modelling.* GARP Publications Series No. 16, 181–97. ICSU/WMO, Geneva.

Arrhenius, S. 1896: On the influence of carbonic acid in the air upon the temperature of the ground. *London, Edinburgh, and Dublin Philosophical Magazine and Journal of Science (fifth series),* **41**, 237–75.

Baldwin, M. P., L. J. Gray, T. J. Dunkerton, K. Hamilton, P. H. Haynes, W. J. Randel, J. R. Holton, M. J. Alexander, I. Hirota, T. Horinouchi, D. B. A. Jones, J. S. Kinnersley, C. Marquardt, K. Sato, and M. Takahashi, 2001: The Quasi-Biennial Oscillation. *Rev. Geophys.*, **39**, 179–229.

Baldwin, M. P., D. W. J. Thompson, E. F. Shuckburgh, W. A. Norton, and N. P. Gillett, 2003: Weather from the stratosphere? *Science*, **301**, 317–19.

Barkstrom, B., E. F. Harrison, G. L. Smith, R. N. Green, J. Kibler, R. Cess, and the ERBE Science Team, 1989: Earth Radiation Budget Experiment (ERBE) archival and April 1985 results. *Bull. Amer. Meteor. Soc.*, **74**, 591–98.

Baytgin, K., and G. Laughlin, 2008: On the dynamical stability of the Solar System. *Astrophys. J.*, **683**, 1207–16.

Betts, A. K., and W. Ridgway, 1989: Climatic equilibrium of the atmospheric convective boundary layer over a tropical ocean. *J. Atmos. Sci.*, **46**, 2621–41.

Bjerknes, J., 1938: Saturated-adiabatic ascent of air through dry-adiabatically descending environment. *Quart. J. Roy. Meteor. Soc.*, **64**, 325–30.

———, 1969: Atmospheric teleconnections from the equatorial Pacific. *Mon. Wea. Rev.*, **97**, 163–72.

Bode, H. W., 1975: *Network analysis and feedback amplifier design*. R. E. Krieger, New York, 577 pp.

Bony, S., R. Colman, V. M. Kattsov, R. P. Allan, C. S. Bretherton, J.-L. Dufresne, A. Hall, S. Hallegatte, M. M. Holland, W. Ingram, D. A. Randall, B. J. Soden, G. Tselioudis, and M. J. Webb, 2006: How well do we understand and evaluate climate change feedback processes? *J. Climate*, **19**, 3445–82.

Cess, R. D., G. L. Potter, J. P. Blanchet, G. J. Boer, S. J. Ghan, J. T. Kiehl, H. Le Treut, Z.-X. Li, X.-Z. Liang, J. F. B. Mitchell, J.-J. Morcrette, D. A. Randall, M. Riches, E. Roeckner, U. Schlese, A. Slingo, K. E. Taylor, W. M. Washington, R. T. Wetherald, and I. Yagai, 1989: Interpretation of cloud-climate feedback as produced by 14 atmospheric general circulation models. *Science*, **245**, 513–16.

Charney, J. G., 1979: *Carbon dioxide and climate: A scientific assessment*. National Academy Press, Washington, D.C., 33 pp.

Charney, J. G., R. G. Fleagle, H. Riehl, V. E. Lally, and D. Q. Wark, 1966: The feasibility of a global observation and analysis experiment. *Bull. Amer. Meteor. Soc.*, **47**, 200–220.

Clement, A. C., R. Burgman, and J. R. Norris, 2009: Observational and model evidence for positive low-level cloud feedback. *Science*, **325**, 460–64.

Cripe, D. G., and D. A. Randall, 2001: Joint variations of temperature and water vapor over the midlatitude continents. *Geophys. Res. Lett.*, **28**, 2613–26.

Crowley, T. J., and G. R. North, 1991: *Paleoclimatology*. Oxford University Press, Oxford, 339 pp.

Donner, L., W. H. Schubert, and R. C. J. Somerville, Eds., 2011: *The development of atmospheric general circulation models: Complexity, synthesis, and computation*. Cambridge University Press, Cambridge, UK, 272 pp.

Drazin, P. G., 1992: *Nonlinear systems*. Cambridge University Press, Cambridge, UK, 317 pp.

Emanuel, K. A., 1994: *Atmospheric convection*. Oxford University Press, Oxford, 580 pp.

———, 2005: *Divine wind: The history and science of hurricanes*. Oxford University Press, Oxford, 296 pp.

Emanuel, K. A., J. D. Neelin, and C. S. Bretherton, 1994: On large-scale circulations in convecting atmospheres. *Quart. J. Roy. Meteor. Soc.*, **120**, 1111–43.

Fleming, J. R., 2009: *The Callendar effect*. American Meteorological Society, Boston, 176 pp.

Fröhlich, C., and J. Lean, 2004: Solar radiative output and its variability: Evidence and mechanisms. *Astron. Astrophys. Rev.*, **12**, 273–320.

Gleick, J., 1987: *Chaos: Making a new science*. Viking Press, New York, 352 pp.

Goody, R. M., and G. D. Robinson, 1951: Radiation in the troposphere and lower stratosphere. *Quart. J. Roy. Meteor. Soc.*, **77**, 151–87.

Hall, Alex, and Syukuro Manabe, 1999: The role of water vapor feedback in unperturbed climate variability and global warming. *J. Climate*, **12**, 2327–46.

Harman, P. M., 1998: *The natural philosophy of James Clerk Maxwell*. Cambridge University Press, Cambridge, UK, 232 pp.

Hartmann, D. L., 1994: *Global physical climatology*. International Geophysics Series, Vol. **56**. Academic Press, San Diego, 411 pp.

Hartmann, D. L., and K. Larson, 2002: An important constraint on tropical cloud-climate feedback. *Geophys. Res. Lett.*, **29**, 1951–54.

Held, I. M., and B. J. Soden, 2000: Water vapor feedback and global warming. *Ann. Rev. Energ. Environ.*, **25**, 441–75.

———, 2006: Robust responses of the hydrological cycle to global warming. *J. Climate*, **19**, 3354–60.

Holton, J. R., 1992: *An introduction to dynamic meteorology*, 3rd ed. International Geophysics Series, Vol. **48**. Academic Press, San Diego, 511 pp.

Imbrie, J., and K. P. Imbrie, 1986: *Ice ages: Solving the mystery*. Harvard University Press, Cambridge, Mass., 224 pp.

Lacis, A. A., G. A. Schmidt, D. Rind, and R. A. Ruedy, 2010: Atmospheric CO_2: Principal control knob governing Earth's temperature. *Science*, **330**, 356–59.

Laskar, J., 1994: Large scale chaos in the Solar System. *Astron. Astrophys.*, **287**, L9–L12.

Laskar, J., and M. Gastineau, 2009: Existence of collisional trajectories of Mercury, Mars, and Venus with the Earth. *Nature*, **459**, 817–19.

Laskar, J., F. Joutel, and P. Robutel, 1993: Stabilization of the Earth's obliquity by the Moon. *Nature*, **361**, 615–17.

Levitus, S., J. I. Antonov, T. P. Boyer, R. A. Locarnini, H. E. Garcia, and A. V. Mishonov, 2009: Global ocean heat content 1955–2008 in light of recently revealed instrumentation problems. *Geophys. Res. Lett.*, **36**, L07608.

Loeb, N. G., B. A. Wielicki, D. R. Doelling, G. L. Smith, D. F. Keyes, S. Kato, N. M. Smith, and T. Wong, 2009: Towards optimal closure of the 'Earth's top-of-atmosphere radiation budget. *J. Climate*, **22**, 748–66.

Lorenz, E. N., 1963: Deterministic non-periodic flow. *J. Atmos. Sci.*, **20**, 130–41.

———, 1969: Three approaches to atmospheric predictability. *Bull. Amer. Meteor. Soc.*, **50**, 345–49.

———, 1995: *The essence of chaos.* University of Washington Press, Seattle, 227 pp.

Lyden-Bell, R. M., S. C. Morris, J. D. Barrow, J. L. Finney, and C. Harper, Eds., 2010: *Water and life: The unique properties of H_2O.* CRC Press, New York, 396 pp.

Manabe, S., and R. T. Wetherald, 1967: Thermal equilibrium of the atmosphere with a given distribution of relative humidity. *J. Atmos. Sci.*, **24**, 241–59.

Marshall, J., and R. A. Plumb, 2008: *Atmosphere, ocean, and climate dynamics.* Elsevier, Amsterdam, 219 pp.

Norris, J. R., and C. B. Leovy, 1994: Interannual variability in stratiform cloudiness and sea surface temperature. *J. Climate*, **7**, 1915–25.

Palmer, T. N., 1993: Extended range atmospheric prediction and the Lorenz model. *Bull. Amer. Meteor. Soc.*, **74**, 49–65.

———, 1999: A nonlinear dynamical perspective on climate prediction. *J. Climate*, **12**, 575–91.

Philander, S. G., 1990: *El Niño, La Niña, and the Southern Oscillation.* Academic Press, New York, 293 pp.

———, 2010: Tilting at connections, from Pole to Equator. *Science*, **328**, 1488–89.

Pierrehumbert, R. T., 1995: Thermostats, radiator fins, and the local runaway greenhouse. *J. Atmos. Sci.*, **52**, 1784–1806.

———, 2010: *Principles of planetary climate.* Cambridge University Press, Cambridge, UK, 652 pp.

———, 2011: Infrared radiation and planetary temperature. *Phys. Today*, **64**, 33–38.

Poincaré, H., 1912: *Science and method.* University of Toronto Libraries, Toronto, 296 pp.

Randall, D. A., M. Khairoutdinov, A. Arakawa, and W. Grabowski, 2003: Breaking the cloud-parameterization deadlock. *Bull. Amer. Meteor. Soc.*, **84**, 1547–64.

Randall, D. A., D.-M. Pan, and P. Ding, 1997: Quasiequilibrium. In *The physics and parameterization of moist atmospheric convection*, R. K. Smith (Ed.), 359–85. Kluwer, Amsterdam.

Randall, D. A., M. E. Schlesinger, V. Galin, V. Meleshko, J.-J. Morcrette, and R. Wetherald, 2006: Cloud feedbacks. In *Frontiers in the science of climate modeling*, J. T. Kiehl and V. Ramanathan (Eds.), 217–50. Cambridge University Press, Cambridge, UK.

Riehl, H., and J. S. Malkus, 1958: On the heat balance in the equatorial trough zone. *Geophysica*, **6**, 503–37.

Schlesinger, M. E., 1989: Quantitative analysis of feedbacks in climate model simulations. In *Understanding climate change*, A. Berger, R. E. Dickinson, and J. W. Kidson (Eds.), Geophysical Monograph 52, IUGG Vol. 7, 177–87. American Geophysical Union, Washington, D.C.

Schneider, S. H., 1972: Cloudiness as a global climatic feedback mechanism: The effects on radiation balance and surface temperature of variations in cloudiness. *J. Atmos. Sci.*, **29**, 1413–22.

Schubert, W. H., P. E. Ciesielski, C. Lu, and R. H. Johnson, 1995: Dynamical adjustment of the trade wind inversion layer. *J. Atmos. Sci.*, **52**, 2941–52.

Seinfeld, J. H., and S. N. Pandis, 1998: *Atmospheric chemistry and physics: From air pollution to climate change.* John Wiley, New York, 1,232 pp.

Shukla, J., and Y. Mintz, 1982: Influence of land-surface evapotranspiration on the Earth's climate. *Science*, **215**, 1498–1501.

Solomon, S., D. Qin, M. Manning, Z. Chen, M. C. Marquis, K. B. Averyt, M. Tignor, and H. L. Miller, Eds., 2007: *Climate change 2007: The physical science basis. Contribution of Working Group I to the Fourth Assessment Report of the Intergovernmental Panel on Climate Change.* Cambridge University Press, Cambridge, UK, 996 pp.

Trenberth, K. E., J. T. Fasullo, and J. Kiehl, 2009: Earth's global energy budget. *Bull. Amer. Meteor. Soc.*, **90**, 311–23.

Vecchi, G. A., and B. J. Soden, 2007: Global warming and the weakening of the tropical circulation. *J. Climate*, **20**, 4316–40.

von Schuckmann, K., F. Gaillard, and P.-Y. Le Traon, 2009: Global hydrographic variability patterns during 2003–2008. *J. Geophys. Res.*, **114**, C09007.

Weart, S. R., 2008: *The discovery of global warming,* rev. and exp. ed. Harvard University Press, Cambridge, Mass., 240 pp.

Wielicki, B. A., B. R. Barkstrom, E. F. Harrison, R. B. Lee, III, G. L. Smith, and J. E. Cooper, 1996: Clouds and the Earth's Radiant Energy System (CERES): An Earth observing system experiment. *Bull. Amer. Meteor. Soc.*, **77**, 853–68.

Wild, M., H. Gilgen, A. Roesch, A. Ohmura, C. N. Long, E. G. Dutton, B. Forgan, A. Kallis, V. Russak, and A. Tsvetkov, 2005: From dimming to brightening: Decadal changes in solar radiation at Earth's surface. *Science*, **308**, 847–50.

Woods, A., 2006: *Medium-range weather prediction: The European approach.* Springer, Berlin, 270 pp.

Xu, K.-M., and K. A. Emanuel, 1989: Is the tropical atmosphere conditionally unstable? *Mon. Wea. Rev.*, **117**, 1471–79.

Zelinka, M. D., and D. L. Hartmann, 2010: Why is longwave cloud feedback positive? *J. Geophys. Res.*, **115**, D16117.

Zhang, C., 2005: The Madden-Julian Oscillation. *Rev. Geophys.*, **43**, RG2003.

Index